Developing to Scale

Developing to Scale

TECHNOLOGY AND THE MAKING OF GLOBAL HEALTH

Heidi Morefield

THE UNIVERSITY OF CHICAGO PRESS
CHICAGO AND LONDON

The University of Chicago Press, Chicago 60637
The University of Chicago Press, Ltd., London
© 2023 by The University of Chicago
All rights reserved. No part of this book may be used or reproduced in any manner whatsoever without written permission, except in the case of brief quotations in critical articles and reviews. For more information, contact the University of Chicago Press, 1427 E. 60th St., Chicago, IL 60637.
Published 2023
Printed in the United States of America

32 31 30 29 28 27 26 25 24 23 1 2 3 4 5

ISBN-13: 978-0-226-82861-9 (cloth)
ISBN-13: 978-0-226-82863-3 (paper)
ISBN-13: 978-0-226-82862-6 (ebook)
DOI: https://doi.org/10.7208/chicago/9780226828626.001.0001

Library of Congress Cataloging-in-Publication Data

Names: Morefield, Heidi, author.
Title: Developing to scale : technology and the making of global health / Heidi Morefield.
Description: Chicago : The University of Chicago Press, 2023. | Includes bibliographical references and index.
Identifiers: LCCN 2023003657 | ISBN 9780226828619 (cloth) | ISBN 9780226828633 (paperback) | ISBN 9780226828626 (e-book)
Subjects: LCSH: Public health—Southern Hemisphere. | Medical technology.
Classification: LCC RA441.5.M67 2023 | DDC 362.10285—dc23/eng/20230202
LC record available at https://lccn.loc.gov/2023003657

⊗ This paper meets the requirements of ANSI/NISO Z39.48-1992 (Permanence of Paper).

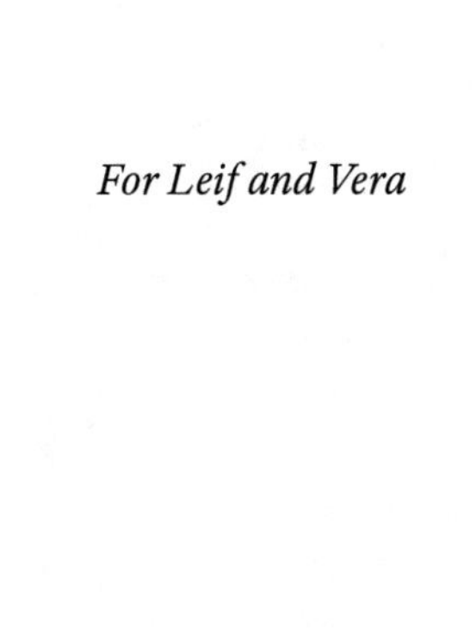

For Leif and Vera

Contents

Introduction 1

CHAPTER ONE
Buddhist Economics · 12

CHAPTER TWO
Small Is Beautiful · 28

CHAPTER THREE
Networking Development · 39

CHAPTER FOUR
Carrots and Sticks · 57

CHAPTER FIVE
Visions of the Future · 74

CHAPTER SIX
The Silver Bullet Boys · 91

CHAPTER SEVEN
Bantu Technology · 113

CHAPTER EIGHT
Scaling Up · 134

EPILOGUE
COVID-19 · 150

Acknowledgments 159

Notes 165

Index 207

Introduction

The roads around Goma are bumpy. Extraordinarily so. The city lies nestled between the shores of Lake Kivu to the south, and Mount Nyiragongo, an active volcano, to the north. It is the capital city of North Kivu province, in the eastern Democratic Republic of the Congo. When I was there in 2010 and 2011, the dried lava and rubble from the volcano's last major explosion, in 2002, still covered many of the roads—making them impassable for most vehicles. Ordinary Congolese citizens could not afford the Toyota Land Cruisers that aid agencies proliferated around the region. On this particular day, I was on my way to Saké, which was twenty-four kilometers to the northwest of Goma. At the time, I worked as a project manager for a United States Agency for International Development (USAID) contractor—a private, for-profit company that fulfilled the United States' international development objectives on behalf of USAID, a government entity. Saké was one of thirty communities across North and South Kivu that hosted the USAID project I managed—aimed at bringing peace and stability to a region plagued by violence. Although it was only twenty-four kilometers away, it might take a resident of Saké all day to reach the hospital in Goma given the condition of the roads. The residents of Saké had therefore proposed a grant to better equip the local health center, and I traveled there to assess the clinic's needs.

The clinic director led me inside, where he pointed out that, aside from basic supplies and some new beds and mattresses, what they needed most were lights. It was very difficult to treat patients after the sun set at 6 p.m., but there was no electric grid and widespread gasoline shortages from the ongoing conflict in the area meant that the generator sat useless, rusting outside. Solar panels were a popular, though costly solution—a good use of the grant money. Before I left, the director led me over to a corner of the clinic, where there was a refrigerator. It looked new. He told me that they'd had it for many months—it had been donated by a major humanitarian relief group to help preserve drugs and vaccines, but with no access to power, it had never worked. He shrugged and said wistfully that that was OK, really, because he'd never had any drugs or vaccines to put in it. "Maybe with the solar panels," he suggested.

I was struck by the stark discrepancy between the clinic's needs and what the staff had been given by aid agencies. Project after project had failed to meet their needs, but nothing about the approach seemed to change. I wanted to know why. Eventually, I began researching the history of global health and international development. I found that my experience aligned with several trends that had taken hold since the 1970s—foreign aid had increasingly been contracted out to private companies like the for-profit entity I worked for, metrics aiming for cost-effectiveness were closely tracked, and aid had become more and more focused on the distribution of small-scale, point-of-use technologies. These commodities encompassed everything from drugs and vaccines to water filters, refrigerators, and off-the-grid power sources. Beginning in the mid-1960s, these commodities became known as "intermediate," or "appropriate," technology—a concept whose definition remained stable over time (technology that was small-scale, locally sourced, labor-intensive, and relatively inexpensive) but open to widespread interpretation when it came to implementation.

The provision of appropriate technologies had largely replaced an earlier form of development prevalent in the 1950s and 1960s. Driven by Walt W. Rostow's theory of modernization, in which countries proceed down a linear path of development to an eventual "takeoff," this version of development oversaw the construction of large-scale, high-capital infrastructure projects. The dams, power plants, highways, and manufacturing facilities were meant to accelerate a country on its path to modernization. By the late 1960s, modernization theory was critiqued for benefiting only the elites within a country (who could afford to plug in to the new power grids and buy the newly manufactured goods) and for the widespread environmental destruction the construction projects wrought.

Appropriate technology is a term used widely, often critically, in the literature on international health and development. It emerged in the early 1970s as a sort of Goldilocks solution—not too traditional and not too modern. While better for the environment than modernization projects, the turn to small-scale, simple technology was not necessarily more sustainable. The late physician and anthropologist Paul Farmer famously reported that a Haitian priest he worked with called it "shit for the poor."[1] It enjoyed a vogue in the 1970s and early 1980s, which was prolonged somewhat in parts of Southern Africa and India, before fading from the vocabulary of development planners and technicians. However, appropriate technology did not really go away. As this book shows, it is fundamental to understanding how the global health enterprise, with its technocentric logic, functions today. Now, appropriate technologies are branded differently—reverse innovation, disruptive technology, and sustainable development are all more modern incarnations that have taken their cue from a much earlier movement.[2]

Given our current context, this is not surprising. Despite recent efforts to broaden the geography of its use, global health remains a concept mostly of the Global North and the West; a set of practices that radiate outward, particularly from the United States. Intervening in international health and development has long been a way of extending the Pax Americana, or American world order, in place roughly from the end of World War II.[3] However, this system, designed to privilege US markets and economic growth at the expense of the rest of the world, began to falter in the mid-1970s. The historian Daniel Rodgers pinned this fracturing of the American world order to the ascendance of a more individualized human nature and the predominance of a capitalist market logic in the last quarter of the twentieth century as the Cold War faded. It has only accelerated since the 2008 global financial crisis, initiated by the recklessness of American banks that were thought to be "too big to fail."[4]

"Disruptive" technologies offer the promise of a quick solution to otherwise intractable problems—a way out of a broken system. In a moment when the overriding narrative is one of widespread breakdown, these sorts of technological fixes can seem very appealing. News media in the United States have been dominated by stories of catastrophe—the COVID-19 pandemic, an impending recession, climate change and environmental degradation, mass shootings, Brexit (the United Kingdom's exit from the European Union), the ongoing wars in Syria, Yemen, and Ukraine, and the foundering of American democratic norms. Other ongoing crises don't receive nearly enough popular attention, although they are no less cataclysmic for the world—like the ceaseless wars and repeated Ebola outbreaks in the Democratic Republic of Congo. Despite the fact that Rodgers shows explicitly how technology furthered the individuation and breakdown he condemns—air conditioners replaced front porches and the need to chat with your neighbors, screens replaced social games—technology is still sold as the solution. Seen in a broader frame, through the lens of the history of infrastructure and the United States' interstate highway system, the historian Jay Sexton has argued that breakdown is, in fact, the regular state of affairs in US history. The stability of the immediate postwar period, until about 1968—the "golden age" of Americana, looked back upon nostalgically by many conservative thinkers—was the anomaly.[5]

It is in the context of breakdown, too, that appropriate technology in its classic form, credited to the economist Ernst Friedrich Schumacher, has been enjoying a bit of a renaissance in his adopted country of England. Writing primarily in the left-leaning *Guardian* newspaper, some commentators have suggested that the solution for everything from the environmental plight to the economic crisis of 2008 could have been mitigated, if not avoided wholesale, had society adopted Schumacher's economic theories. In late 2011, the columnist Madeleine Bunting celebrated *Small Is Beautiful: Economics as If People Mattered* (1973), Schumacher's best-known work, writing that "it is chilling that so many thinkers, politicians, and academics have signed up to the deadening consensus of globalisation."[6] Inspired by the then-apparent "collapse" of the euro, the columnist Andrew Simms wrote another piece a few days later about how Schumacher's work wasn't so much about smallness in and of itself but rather appropriateness of scale: "Every neighbourhood might, therefore, have its own bakery, but not a factory making trains." The European Union had failed to balance this equation of scale. "Too big to fail" was bitterly ironic for Schumacher's acolytes, while "the European Union . . . is collapsing partly because its super-size currency is chronically incapable of meeting the needs of such a diverse range of economies."[7] They foresaw Brexit as the result.

∴

Appropriate technology is the product of the deep economic and political unrest of an earlier generation. To trace a brief genealogy of the concept, it is necessary to understand the context in which the movement took hold. The decades immediately following World War II were unusually economically stable. This was in spite of major political gains within the civil rights movement and for many women through the work of second-wave feminists in the United States. These wins were far from comprehensive—second-wave feminists have been thoroughly criticized for focusing on white, middle-class, straight women to the exclusion of women of color, queer women, and those from lower classes, all while the fight to implement hard-won civil rights largely remained. They were also not the reason that many older white Americans look back through rose-colored glasses on this period, imagining scenes reminiscent of *Leave It to Beaver*. The decades of the mid-twentieth century were a time of American economic boom and widespread cultural hegemony. The white middle class was thriving, whatever the discontents of its marginalized groups.

Outside of the continental United States, the late 1940s through the late 1960s was a period of widespread decolonization internationally, particularly in Africa and Asia.[8] Postcolonial leaders informed the processes of state building that were taking shape across wide swaths of the world with critical theories by Aimé Césaire, Albert Memmi, and Frantz Fanon.[9] Old systems, developed and entrenched through colonial rule, had to be broken and rebuilt in order to ensure liberation. The process of rebuilding focused, unsurprisingly, on science and technology—not just their acquisition from elsewhere, perpetuating cycles of dependency, but the fostering of indigenous manufacturing and industrial capacity. The historian Abena Osseo-Asare has written about how the desire for scientific and technological equity in the immediate postcolonial period can simultaneously be understood as an "assertion of sovereignty" and, in their reliance on imported laboratory equipment, a "sign of dependence."[10] Many postcolonial states embraced socialism or nationalism as utopian means of procuring, fostering, and disseminating science and technology, but as the historian Samuel Moyn has shown, as these political ideals crumbled in the 1970s, the human rights discourse that replaced them focused on individual rather than communal justice.[11] On a global scale, Moyn contended, the 1970s represented a shift from a rhetoric of "equality" to one of "sufficiency," even with the rise of the New International Economic Order.[12] Appropriate technology helped hasten these ends, as it was embraced by donor countries as a way out of providing equitable access to science and technology.

The postcolonial theorist Achille Mbembe has argued that discourses about Africa are trapped in Western tropes about the continent's "lack"—its possessions seen as inferior to those of the West. Postcolonial leaders worked tirelessly to overcome this in the 1960s and early 1970s with a thoroughly modern vision for the continent's future.[13] Leaders of the Non-Aligned Movement formed in 1961, including Jawaharlal Nehru of India, Kwame Nkrumah of Ghana, Sukarno of Indonesia, Josip Broz Tito of Yugoslavia, and Gamal Abdel Nasser of Egypt, aimed to achieve economic development through mutual cooperation. They lobbied at venues like the UN Conference on Trade and Development (UNCTAD), founded in 1964, for better terms of trade and the transfer of technology.[14] The United States, which under the Kennedy administration declared the 1960s the "Development Decade," was instead intent on delivering foreign aid via US manufacturing and other business interests.

In 1968, discontent with the Vietnam War after the Tet Offensive and the increasing critique of modernization-driven development meant the US Congress lost much of its public support for foreign aid—USAID's budget was cut 26 percent in inflation-adjusted dollars between fiscal years 1968 and 1969.[15] The 1973 oil shocks, on top of the already enormous costs of the Vietnam War, then created a fiscal crisis within the United States that led to even more severe cutbacks in the foreign aid budget—a further 16 percent reduction in constant dollars, or 0.6 percent of the overall budget.[16] In the midst of the Cold War, the United States could not abandon its commitments in the developing world, but it did have to find a cheaper way of engaging in foreign aid and technical assistance.[17]

Appropriate technology, based on Schumacher's concept, emerged as a potential way out of the fracas. To its proponents, it made intuitive sense. The historian of technology Carroll Pursell, for example, explained Schumacher's philosophy as follows: "Technologies sent to Third World areas should be carefully chosen to create workplaces that were located where people lived, that were cheap enough for ordinary people to acquire, and used relatively simple techniques and local materials to make things for local use."[18] For Pursell, these ideals were "as relevant to overdeveloped economies as [they were] to underdeveloped economies." In practice, appropriate technology programs often focused on principles that would later be attributed to neoliberalism and the rise of the administrations of Ronald Reagan and Margaret Thatcher: privatization, cost-effectiveness, and austerity.[19] These principles were glossed and understood as more social, ethical, and commonsense solutions—words like *thrift* carried a more moral valence than the severe *austerity*—but they effectively meant the same thing.

Yet appropriate technology was often cast as the antithesis of neoliberalism. Common accounts portray a hard break between the benevolent 1970s appropriate technology programs and the advent of Reaganism. Pursell's work on the history of appropriate technology focused mainly on the agricultural sector and the impact it had on the export of US industrial technology, which led him to conclude that the appropriate technology concept did not survive the 1970s.[20] Langdon Winner, another historian of technology, similarly proclaimed that "the demise of appropriate technology as a living social movement can be dated almost to the hour and minute. It occurred early in the evening of Tuesday, November 4, 1980, when it became clear that Ronald Reagan had been elected President of the United States."[21] His take on the appropriate technology concept, more critical than Pursell's, linked its rise to enthusiasm for alternative energy and a cynical retreat, beginning in 1968, from more broad-based political organizing. For Winner, appropriate technology was done in by postindustrial capitalism.

With the benefit of a wider lens and a longer range of hindsight on the supposed death of appropriate technology, we can see that the 1980s were not its end, but a period in which the movement regrouped and rebranded in part through initiatives occurring outside of the United States—especially in the developing world. In critiques of technological determinism, historians and science and technology studies scholars have shown how the materiality of technology—exploring how specific technologies function and the assumptions underlying their design and use—allowed for unexpected claims for sovereignty and agency from actors in the Global South.[22] Yet until relatively recently, most scholarship in the field focused on large-scale high-technology systems. The historians of Cold War biomedicine have likewise tended to focus on large-scale multilateral collaborations such as family planning and the malaria and smallpox eradication campaigns.[23]

Many appropriate technologies billed as "new" solutions were not especially innovative. The same goes with most "disruptive" technologies today. Andrew Russell and Lee Vinsel have been particularly outspoken in criticizing technology studies' preoccupation with innovation at the expense of the often more important project of technological maintenance.[24] Within the turn toward considering less innovative, more mundane everyday technologies is the work of David Edgerton, who argues for different ways in which mundane technologies can produce the "shock of the old," and David Arnold, who considers the politics inherent in everyday Indian technologies such as the bicycle.[25] Clapperton Mavhunga has looked at everyday innovation from an African perspective, using the example of poaching in Zimbabwe to argue that—in contrast to development policies which are "*for them* but not *with them*"—as ordinary people engage with technology they are not merely users but, in assigning incoming objects "new meanings and purposes," designers in their own right.[26] This holds true in the case of appropriate technology, although the battle over defining *appropriate* had very real implications for development policy.

Critical approaches to the international health and development sector have also flourished, noting the lack of consideration for local context and, above all, the people subjected to these programs as "beneficiaries."[27] Although the span of this literature is very attentive to questions of interventionism, neocolonialism, and bioethics in international health programs, little attention is paid to the specific technologies that allow these programs to function—whether DDT, the bifurcated needle, or contraceptives.[28] Tracing these programs through the 1970s and into the 1980s, over the bumpy terrain of USAID budget cuts and through the supposed neoliberal turn of the early 1980s, the history of appropriate technology, and in particular how it was deployed through American foreign aid, challenges the common narrative chronology within the history of global health, wherein commitments to broad-based primary health care initiatives at the World Health Organization—and the 1978 Alma-Ata Declaration that affirmed them—fell victim to Reaganism and a World Bank bent on measuring everything.[29] Rather, primary health care was stillborn—the factors that led to its quick demise were firmly entrenched years before Alma-Ata.

Appropriate technology was adopted as an instrumental part of community-based primary health care delivery. The technologies, like vaccines, latrines, and essential drugs, were meant to complement broad, affordable access to social medicine in the clinic. But they were also seen as cost-cutting measures, relative to big modernization projects, and to some policy makers they obviated the need for comprehensive primary care. Just a year after Alma-Ata, experts meeting to discuss how to implement primary health care decided that it would be impractical to implement all at once and that an "interim strategy" was more feasible in the immediate term. It's not surprising that the strategy focused on measures that were most cost-effective, or that the experts relied on appropriate technologies—these things went hand in hand. Infant growth monitoring (usually with simple paper tape measures), oral rehydration (with inexpensive packets of a salt and sugar solution), breastfeeding, and immunization campaigns did not need underlying infrastructure or structural changes to work. Distributing tape measures and oral rehydration packets and training community health workers was cheaper than setting up rural clinics and sustaining them for the long term. Taken together, these interventions became known by the acronym GOBI (for growth monitoring, oral rehydration, breastfeeding, and immunization), the hallmark of "selective" primary health care. Divorced from social medicine, appropriate technology enabled the provision of international health to center around the distribution of discrete commodities.

Another common critique is that through public-private partnerships, which are common arrangements in international development today whereby government aid agencies and nongovernmental organizations (NGOs) work with the for-profit private sector, capitalist profit motives have corrupted humanitarian endeavors. Yet economic imperatives have been a major impetus for US foreign aid since President Truman's Point Four program, a technical assistance program for developing countries announced in Truman's inaugural address in 1949 as the Cold War progressed. These American economic imperatives tend to get written out of the literature on development and modernization, with a focus instead on the political incentives the United States had to win over developing countries to its sphere of influence in the global Cold War.[30] More recently, Peter Redfield, looking at twenty-first-century global health technologies—"ingenious, small-scale gadgets"—such as the LifeStraw (a point-of-use water filter for individuals), argued that, although these new technologies are very much in alignment with market prerogatives, they nevertheless seek to do good.[31] This is particularly true as the balance of technological assistance programs has shifted, as has the human rights discourse, away from collective infrastructure projects, like communal water pumps, toward individual consumables, like the LifeStraw, over the course of the later decades of the twentieth century.[32] Appropriate technology, though ostensibly for humanitarian aid, has been subject to the same economic incentives as other commodity flows to and from the Global South.[33]

∴

Just as the idea of appropriate technology sprawled across different fields and networks, transforming subtly with each twist and turn, so, too, did my research. This book is based on visits to sixteen different archives and personal paper collections and more than twenty oral history interviews over four countries. I traced the records of private organizations, like PATH and the Gates Foundation, that do not open their archives to the public, through the files of their public (or, more public) donors and partners—including USAID, the World Health Organization, the United Nations, the World Bank, and the Ford Foundation. I interviewed their founders, their former and current contractors and staff, and the people who work in their field offices. I also attempted to interview the people who were the intended beneficiaries of these types of global health programs in South Africa and Zimbabwe, although this was a difficult undertaking and generated mixed results. Many people I spoke with were still dependent on donor money for their livelihoods, so they were reluctant to speak out on the record. When they were candid with me, I have had to carefully consider how much of what they said to include, even when anonymizing sources, so as not to jeopardize their careers or future grants. At the same time, I found the people I interviewed at PATH and at Gates to be very aware of common criticisms and self-reflective in how they addressed them. As a former humanitarian fieldworker myself, I know we can be some of the harshest critics of the moral pretensions, ambiguities, pernicious effects, and frequent failures of our own work.

This is a history of the technological fix in global health, for better and for worse. It is not a story with a villain. Rather, my goal is to show the ways in which an idea—appropriate technology—has been assembled, appropriated, disseminated, transformed, discarded, reconstituted, and, eventually, sold and made anew. The actors *doing* this thought work, and the actors acting on and reacting to their words, varied rather wildly—from Stewart Brand to Robert Mugabe; Union College engineers to the United Nations; Jimmy Carter to Bill Gates. Mostly, they were the mid-level bureaucrats, project managers, development technicians, local experts, and men and, especially, women who were meant to use the technologies, putting theory into practice. There are myriad other ways in which both the concept and the term *appropriate technology* have been leveraged in other fields and by other actors, which are not the subject of this book. I therefore make no claim to present a comprehensive history or genealogy of appropriate technology. Rather, in these pages I attempt to capture the breadth of ways and spaces in which appropriate technology permeated international development and shaped the field and practice of global health today.

This book is organized more or less chronologically. In chapter 1, I follow Schumacher's intellectual development from his early inspiration—notably his visits to Burma and India—through to his articulation of appropriate technology as a concept. He did not assemble his system of thought in a vacuum, but rather he borrowed many ideas from his interpretation of the Buddhist faith, Gandhism, Hindu economics, and fellow exiled intellectuals, which are explored in chapter 2. Chapter 3 explores the networks created with the goal of disseminating small-scale technologies and how they were put into the service of different social causes, including feminism and environmentalism, through the resonances of textual catalogs. Although Schumacher had intended his theories to be for the developing world, many of his followers took up his ideas, and concurrently developed their own, in the United States and United Kingdom. Through publications like the *Whole Earth Catalog* and the *Village Technology Handbook*, this preinternet network of engineers and technicians helped put the idea of appropriate technology into practice, according to their own, eventually techno-utopian and libertarian, interpretations.

Who defined what appropriate technology was? I consider the competing visions of appropriate technology through the many bilateral (chapter 4) and multilateral (chapter 5) meetings that took place during the 1970s. The West's relatively constrained version of appropriate technology (cheap, simple, off-patent) eventually won out and would be leveraged to fend off demands from the developing world for scientific and technological equity and maintain the status quo. But as USAID began to contract out development projects to a new flock of nongovernmental organizations in the late 1970s, as chapter 6 shows, NGOs were able to redefine the concept to fit their own initiatives. In international health, PATH developed its own novel small-scale devices that challenged perceptions of appropriate technology as old or improvised. This had particular consequences for women, who were most often the target recipients of international health campaigns.

In chapter 7, I follow the appropriate technology concept as it was implemented in the service of liberation struggles in Zimbabwe and apartheid South Africa in the 1980s. Insulated from many foreign aid programs for much of the 1970s as a result of the civil war in Zimbabwe and the apartheid regime in South Africa, appropriate technology enjoyed a somewhat delayed bloom in Southern Africa. Those programs, too, tended to focus on women, in part because women in newly independent Zimbabwe, like Joice Mujuru, held considerable power. Women were also important leaders of South Africa's National Progressive Primary Health Care Network, which was instrumental in designing the country's Health Act during the postapartheid transition. The residue of appropriate technology was difficult to scrub off of global health programs. Chapter 8 considers how PATH's model has transformed the twenty-first century's tech-driven global philanthropy. The immense scale of the model (as currently interpreted) helps us understand how and why technological solutions have come to dominate global health at the expense of longer-term, more equitable approaches to linking technology, health, and development.

Finally, I close with a brief epilogue reflecting on how this model of global health—focused on the distribution of small-scale, inexpensive, point-of-use technologies—has fared in the COVID-19 pandemic. From ventilators and masks to rapid test kits and vaccines, our response to the pandemic has largely mimicked approaches of the past while also exposing the deep-seated structural failures that leave poorer communities at much greater risk. My hope is that this book will help set us on a better course for the future.

[CHAPTER ONE]

Buddhist Economics

Ernst Schumacher sat at his typewriter in Prome Court and took stock of his frustrations. The large, pillared balconies of the colonial-era apartment complex looked out across the palm trees lining Pyay Road at the teak-framed Pegu Club, once the storied haunt of British officials in Burma's capital, Rangoon.[1] The man, who would go on to become a foundational figure in the appropriate technology movement, had been there since early January—part of a three-month stint as an economic adviser to the Burmese government under the auspices of the United Nations—and that day, March 29, 1955, was to be his second to last on the job.[2] Since Burma's independence, Prome Court had housed important government offices. His task was to assess the development plans that a group of American consultants had prepared for the new government, but it hadn't gone well.[3] As he sat, he composed a letter to U Thant, then the executive secretary of the Economic and Social Board and close confidant to Burma's prime minister U Nu.

Marked "Strictly Personal & Confidential," Schumacher grumbled to Thant that he had been unable to fulfill his UN job description. He quoted it at length: "The expert will work in close association with the Economic and Social Board and the Ministry of Finance and Revenue. His will not be a routine assignment, but one of advising Government at the highest level on subjects of far-reaching importance."[4] Schumacher attributed his failure to Thant's apparent decision to politely ignore him—"no 'close association' has developed; no requests for advice have come to me from any of the members of the Economic and Social Board; and my various memoranda which, whatever their shortcomings, certainly did deal with subjects of far-reaching importance, have so far failed to evoke any demands for discussion or further elucidation. All I can hope for now is that there may possibly be some 'delayed action' effect—that my writings will be found worthy of careful consideration by 'Government at the highest level' which I had been meant to advise."[5]

Thant didn't seem to think much of Schumacher's role. A skilled and important diplomat in his own right, he had taken over the directorship of Burma's Economic and Social Board the day after Schumacher arrived. A devout Buddhist, Thant was frequently described as "deceptively calm," "extraordinarily polite," and a "quiet, patient negotiator."[6] He was born into a wealthy Burmese family and spent many years as a schoolteacher before accepting a post in the first government of independent Burma at the request of his friend from the University College in Rangoon who would become Burma's first premier, U Nu. Both Nu and Thant had been student leaders of the nationalist movement against British colonization in the 1930s and against the Japanese occupation in the mid-1940s. Burma achieved independence from the British in 1948 after the Japanese had surrendered. U Thant started as press director, and was later minister of information, under Nu's administration.

Schumacher closed his letter graciously by thanking Thant for his "unfailing courtesy and kindness." Privately, he blamed him for his modest influence in Rangoon. In his report to the resident representative of the UN Technical Assistance Board, he attributed his frustrations in Rangoon to the "change in personnel" in the directorship of the Economic and Social Board at the beginning of his visit.[7] Despite the tensions, Thant remained friendly in his exercise of diplomatic formalities. He hosted a luncheon in Schumacher's honor the next day at the Union of Burma Club, thanking him for his service to the country.[8] But when the United Nations proposed a return visit by Schumacher the following year, the government declined.[9]

The problem was that Schumacher's concept of what Burma needed flew in the face of the dominant theories of economic development at the time. U Nu's government had been advised by the Washington, DC–based firm Robert R. Nathan Associates, whose more than one-thousand-page plan for the country's economic development included large-scale infrastructure projects like the construction of dams, railways, and highways; hospitals; irrigation projects; coal and hydroelectric power plants; as well as massive industrial development and the exploitation of natural resources.[10] This was in keeping with economist W. W. Rostow's modernization theory, which purported that such large-scale, high-capital development would enable the "takeoff" and rapid growth of developing country economies.[11]

Schumacher didn't think that strategy fit Burma. In the first month he'd been there he'd grown very fond of the country. "The people really are delightful," he wrote to his wife, Anna Maria. "Everything I had heard about their charms and cheerfulness proves to be true . . . There is an innocence here which I have never seen before . . . In their gay dances and with their dignified and composed manners, they are lovable; and one really wants to help them, if one but knew how. Even some of the Americans here say: 'How can we help them, when they are much happier and much nicer than we are ourselves?'"[12] Schumacher was troubled by how to help without harming the Burmese culture, which he appeared to romanticize in a naively racist way. Despite what the national income statistics said, he did not see the average Burmese person as poor, because in his view their simple standard of living made them want for very little. He worried somewhat patronizingly that the intrusion of Western development planners would create poverty and want where it hadn't been before, writing to his wife, "The whole Orient is coming out in Western spots."[13]

While in Rangoon, Schumacher fell in love with Buddhism. He spent his free time studying with local masters and regularly attended a Buddhist monastery to practice meditation.[14] For Schumacher, Buddhism was the undergirding of the Burmese peoples' simple lifestyle; its best defense against Western materialism.[15] His folders from the trip to Burma contain handwritten documents outlining the principles of the Buddhist faith, with a focus on living simply.[16] Burma had a long history, prior to colonization, of a relatively egalitarian society organized into self-governing villages.[17] Schumacher wanted to preserve the nation's character, not set it on the path to Westernization. He didn't see large-scale infrastructure like power plants and highways as universal needs. Rather, he proposed that the choice of technology for development depended significantly on the context in which it was to be used, and in a sort of religious determinism, he defined Burma by what he took to be its most standout feature: its dominant Buddhist faith.

For Schumacher, it followed naturally that the country's economic system should fit its cultural context, but this wasn't necessarily a priority to U Nu. In fact, Burma's government had already endorsed the Nathan & Associates plan. In 1954, the Burmese Economic and Social Board published *Pyidawtha—The New Burma* under its name, although it was a marketing publication that Nathan & Associates had prepared for them. It laid out the development plans for Burma in plain language, to win the support and enthusiasm of everyday citizens. It was ambitious and grand. The health plan, for example, included provisions for a pharmaceutical manufacturing plant, while the "industrial program" planned for the manufacture of everything from pulp and paper to steel products and tires.[18] Nu's vision of the new Burma was thoroughly modern, whatever the religious affiliation of its majority population.

The Cold War was raging, and both the American and Soviet factions were keen to win Burma over to their vision of development. The entrenched American economists advising the government promised economic growth on a massive scale—the Nathan & Associates plan aimed to grow the economy by 63 percent by 1960.[19] The economists hoped that industrial modernization would bring Burma firmly into capitalist world markets and the Unites States' sphere of influence. U Nu, like many Third World leaders, wanted modern technology for his country and was able to use this dynamic to his advantage.[20] The United States was alarmed, several months later, when Nu made a visit to the Soviet Union and cut a deal to trade Burma's surplus rice for Russian technical aid.[21] While Nu was trying to secure the country's economic future—with assistance from wherever he could get it—Schumacher's vision of small-scale, incremental development that largely preserved the traditional Burmese was of life was not appealing.

Without much of an official audience for his ideas, Schumacher took to writing them up privately. He circulated the paper he produced, "Economics in a Buddhist Country," among a small group within the Burmese government he thought would be sympathetic, although they weren't impressed.[22] Frustrated, he planned to give a lecture expressing his ideas for Buddhist economics at Rangoon University. The organizers who had invited him got word of his intentions and shut the talk down, so he "got a typist to make copies and passed them around."[23] In the paper, he makes the case for a wholly new form of economics. "What today is looked upon as *the* science of Economics is based on *one* particular outlook on life, on one only, the outlook of the Materialist. Every concept of Economics is rooted in this outlook," he wrote.[24] He went on:

> This one-sidedness of Economics is surprising and indeed abnormal. Yet it is understandable all the same. For two reasons: first, because *up to a point,* as I have said, everybody is inescapably concerned with material, economic things, if, indeed, he wants to live in a becoming way. *Up to a point* therefore, Economics is about life as such, irrespective of any ideas of meaning and purpose. The second reason is of an altogether different kind: Economics as a science has arisen only in the West and at a time when Western Materialism ruled supreme throughout the world. Non-materialists have been too weak, so far, to think these matters out from their own point of view. And it is one aspect of their continuing weakness that they have thoughtlessly and all too easily accepted the spurious claim of Western Economics to be the only possible body of economic thought, to be final, and objective, applicable to all men at all times.[25]

Certainly, framing the Nu government as "weak" and "thoughtless" was perhaps not the best way to win them over to his cause.

He went on to critique the ethos of "progress without limit" that characterized mainstream economics. "Burma must become a progressive nation, so that her people not only live better in 1960, but look forward to continued improvement, without limit," Schumacher quoted from the thousand-page plan Nathan & Associates he had been brought to Burma to evaluate, before protesting that such progress was incompatible with Buddhism or Christianity, "or with anything the Great Teachers of mankind have proclaimed."[26] For Schumacher, limitless growth was "compatible only with the most naked form of Materialism."[27] Economics, as a discipline, was not value-neutral—it reflected the moral and philosophical interests of Western capitalism.

What was the alternative to materialist, Western economics? Schumacher put forward Mahatma Gandhi's economics, which he viewed as compatible with Hinduism and Buddhism, as a model. This economics relied on the principles of *swadeshi*, to buy things locally, and *khaddar*, to make things oneself. He gave the example of freight rates to illustrate his point. In most Western countries, the cost of shipping goods on a railway, by inland waterways, or by truck tapers off the longer the distance to be shipped. For a traditional economist, Schumacher argued, this "encourages long-distance transport, promote[s] large-scale, specialized production, and thus leads to an 'optimum use of resource.'" However, he urged, it is also furthering the materialist values implicit in that long-distance, large-scale production. By contrast, a Gandhian economist would, Schumacher surmised, give very different advice: "Local, short-distance transportation should receive every encouragement; but long hauls should be discouraged because they would promote urbanization, specialization beyond the point of human integrity, the growth of a rootless proletariat, in short, a most undesirable and uneconomic way of life."[28]

Rather than follow the plan of the American economic advisers in Burma, therefore, Schumacher exhorted his Burmese colleagues to find their own "Middle Way" of economic development, if they wanted "to remain Buddhists."[29] He envisioned an economics based on renewable resources that gave the people just enough, but not too much. "We can distinguish three economic conditions," he wrote, "misery, sufficiency, and surfeit. Of these, two are bad—for a person, a family, or a nation—and only one, sufficiency, is good. Economic 'progress' is good only to the point of sufficiency; beyond that, it is evil, destructive, uneconomic."[30]

Schumacher's critique of economics as a Western discipline didn't sway the postcolonial administration. Why should their citizens be denied the same material advantages that others enjoyed in the West? Moreover, as the debate between capitalism and communism persisted in the larger

Cold War, it was difficult to know what to make of Schumacher's theories. On the one hand, the values of communal well-being he espoused were more in line with socialist rhetoric. On the other, although Schumacher questioned the scale of the Nathan & Associates plan, he left the broader capitalist framework—of trade and industry in the hands of private individuals—regulated but intact. There was still a role for private businesses in supplying development technologies. While in Burma, Schumacher was lobbied by British export outfits who were eager to expand into new markets, like Burma's. The Massey-Harris-Ferguson company of Coventry, which sold British-made tractors, asked him to focus on the impact agricultural mechanization could make in the country.[31] The general sales manager wrote, "My own experience of the Burmese Government purchasing at the present time is that they pay little or no attention to the suitability of the equipment they purchase, and there are no sour grapes about this as we have obtained as much of the business as we can expect."[32] Market expansion was, for British and American companies and their governments, a major benefit of foreign aid.

Schumacher's experience in Burma stayed with him. He would later refer to the country as an example in many of his more popular writings—most famously, in a chapter of his 1973 book *Small Is Beautiful: Economics as If People Mattered*, titled "Buddhist Economics." He had originally published that essay in 1966, as a chapter in *Asia: A Handbook*.[33] In it, he drew explicitly on Burma and the Right Livelihood principle of the Buddhist Eightfold Path. He claimed that since there was a principle of Right Livelihood, there must be such thing as Buddhist economics. Quoting from *Pyidawtha—The New Burma*, which stated that "we can blend successfully the religious and spiritual values of our heritage with the benefits of modern technology," he argued that there was a disconnect between wanting to remain faithful to one's Buddhist heritage and hiring modern economists and development experts from the West. "No one seems to think that a Buddhist way of life would call for Buddhist economics, just as the modern materialist way of life has brought forth modern economics."[34] Rather than a "new Burma," he was calling for a new economics.

∵

Ernst Friedrich Schumacher, known to friends and family as "Fritz," was born in Bonn in 1911. In 1930, he won a Rhodes scholarship to study economics at New College in Oxford, and he completed his studies as an exchange student at Columbia University in New York. Vehemently opposed to the Nazis taking hold in his native Germany, he returned to England after completing his undergraduate degree. As the war broke out, he was labeled an "enemy alien" within the United Kingdom. David Astor, a friend from Oxford, secured a place for him as a farmhand on his uncle Robert Brand's estate, Eydon Hall, in Northamptonshire.[35] In May 1940, under suspicion of being German, he was arrested and spent ten weeks in the Prees Heath internment camp in Shropshire before influential friends, including the Astors and Brand, secured his release back to Eydon Hall.[36]

Between his toil on the farm, he continued to write and publish papers on economics. One, a plan for an international currency clearing office titled "Free Access to Trade," was sent via David Astor and Robert Brand to John Maynard Keynes. Keynes saw potential in young Schumacher. As he took him under his wing, he became a key influence on Schumacher's economic philosophy and introduced him to important figures in the London economics scene, including R. N. Rosenstein Rodan, who directed the Royal Institute of International Affairs at Chatham House.[37] Buoyed by Keynes's support, Schumacher found a position at the Oxford Institute of Statistics and moved back to New College in March 1942.

Schumacher believed firmly in the importance of state intervention into the market for the social good. He continued to press his scheme for an international currency clearinghouse. He was a regular speaker at Chatham House, and his plan eventually made it into the hands of government officials, including the chancellor of the Exchequer, Sir Stafford Cripps. After meeting with Cripps and several other Treasury officials, Schumacher decided to publish the paper formally—it appeared in the May 1943 edition of *Economica*.[38] Keynes had incorporated many of Schumacher's ideas a month earlier in his "Proposals for an International Clearing Union," a government white paper that Keynes later put forward at the Bretton Woods conference.[39] An oft-repeated story among Schumacher's followers is that Lord Keynes himself had, on his deathbed, told Sir Wilfred Eady of the Treasury, "If my mantle is to fall on anyone, it could only be Otto Clarke or Fritz Schumacher. Otto Clarke can do anything with figures, but Schumacher can make them sing."[40] Schumacher himself wrote Keynes's obituary for the *Times*.[41] Yet many of his later writings can also be read as a response to what Schumacher saw as the failures of Keynesianism, particularly in the developing world.

By all appearances, Schumacher was a thoroughly traditional economist for the first twenty-five years of his career. He flirted with Marxism and socialism in the early 1940s, believing firmly in "state planning, large-scale state monopolies, mass production and standardization."[42] But he didn't embrace Marx fully. He had a hard time reconciling state planning with individual freedom, although he supported the social democratic program of the Labour Party in the immediate postwar period in Britain, including the nationalization of the coal industry.

Burma changed him. It was as much a spiritual awakening as a professional one. Through his study of Buddhism and his encounters in Burma, Schumacher came to believe that humankind needed a higher meaning and purpose in life than simply fulfilling quotidian material concerns. Although he didn't leave much record of why this transformation came to be or exactly how his thinking evolved, Schumacher came to reject the atheism of his young adult life.[43] His faith in Buddhist principles led him to accept that there was likely truth in Western religious traditions as well. Quoting the advice of Gandhi to study his "ancestral religion" and "serve it by purifying its defects," he undertook a rigorous study of Christianity upon his return home.[44] He would, many years later, formally convert to Roman Catholicism, inspired by the social and political ideals of Catholic thinkers like G. K. Chesterton and Hilaire Belloc.[45]

Schumacher's newfound spirituality led him to question many of his beliefs about economics and the wider world. He drew a contrast between what he called *Homo sapiens*, the scientific man, and *Homo viator*, the created man. The former had no purpose in life—he was an "accident" of evolution; part of a deterministic, mechanistic system within which he had no freedom, and could only "aspire to raise himself to power."[46] The latter, *Homo viator*, was created with a purpose—to "develop within himself higher powers of consciousness," which distinguished humans from animals and made him free.[47] This freedom gave *Homo viator* responsibility for the choices he made about what to do in his life. Schumacher did not believe that religion and science were naturally in conflict because they were separate: religion on a higher, spiritual plane; science on a lower, material one. However, if scientists and researchers claimed 100 percent certainty over the happenings of the material world, as opposed to just 99.99 percent, then, he wrote, "[science] steps outside its proper knowledge and is in total conflict with religion."[48]

Science, unchecked by doubt and motivated by a desire to control the world, was, for Schumacher, "the extreme formulation of the Hubris of *Homo sapiens*, who wants to become god. As the serpent said, 'And ye shall be as gods.'"[49] As Genesis 3:5 made clear, although humans gained God's knowledge of good and evil, the forbidden fruit conveyed neither God's power nor God's wisdom. But knowing evil was key to human nature. Schumacher drew a connection between the lack of spirituality and meaning in life and the rise of hateful dogma. He believed Nazism had filled a vacuum, an ideology that gave people purpose and a chance at God-like power, while the German intelligentsia had shirked their responsibility to stand up for truth and higher ideals in the early days of Hitler's rise. *Homo sapiens*, scientific man, suffered from a crisis of personal responsibility, he thought.

Schumacher blamed a "devilish trio" for what he saw as the modern refusal to accept or acknowledge individual responsibility and the need for personal growth: Freud, Marx, and Einstein. Freud had shifted responsibility from the individual to sexual frustration and had reconceptualized human relationships according to a logic of self-fulfillment, without consideration for others. Marx had shifted responsibility from the individual to the bourgeoisie and had channeled personal concern into class hatred. Einstein, meanwhile, had replaced "absolute standards" of good and evil with the idea that everything was relative—providing the "perfect excuse for avoiding personal responsibility, or upholding or striving for the good and the beautiful."[50] Where *Homo viator* pursued politics, economics, and art as responsible servants helping humankind to reach higher planes of existence, *Homo sapiens* pursued them only out of individual greed, lust, and a Machiavellian desire for power.

Although he continued to admire Marx's writing, after Burma, Schumacher parted ways with what he thought of as Marx's excessively materialistic analysis. Without a higher spiritual purpose, Marx's atheism led to a "petty bourgeois hatred of what one could not understand."[51] The dignity of work was extraordinarily important, Schumacher agreed, but he could not abide the class hatred, materialistic interpretation of history, and insistence on violence that Marx espoused. He wrote that Marxist analysis engaged in "a hateful and mean kind of depth analysis . . . similar to Freudian psychology. Everything has a sinister 'deeper' meaning. This is the final degradation of the intellect. 'Double talk,' Utopia of a classless society which forgets all about sin, is thus burdened with the previous undermining of all standards of decency . . . The necessary critique of capitalism and bourgeois degradation proceeded through means that were themselves degraded."[52]

He was also critical of what he called the religion of economics. Rooted in materialism and the "false belief that it was a science," capitalist economics promoted greed. "Economics is typically the product of the bourgeois spirit which, recognizing nothing beyond the visible world, is solely interested in *manipulating the world* 'for gain,'" he wrote. This led to a demeaning of higher ideals, to "contempt' of people, and to the abuse of nature."[53] The solution was to reorient economics—as well as politics, medicine, agriculture, industry—toward the Buddhist principle of *ahimsa*, or nonviolence. This took inner development, of the spirit, as a necessary precondition for economic development.

Disenchanted with the traditional economic creed, he came to the conclusion that large-scale planning and standardization—whether according to a free market capitalist logic or a centrally planned communist one—marked the opposite of development progress. Economics as a field should be molded according to more humane principles, he thought, but was nevertheless an important tool in satisfying the reasonable material wants of the global population.[54] Reasonable, for Schumacher, was the equivalent of a Buddhist Middle Way; it was sufficiency, not misery or surfeit. He embraced his role as an economic expert, but his advice diverged markedly from the mode of "capitalist development" that positioned itself in the mid-twentieth century, in the historian Timothy Mitchell's words, as "natural resources versus technology, bodies versus hygiene, men versus machines."[55] Mitchell pointed out that capitalism, and the science and technology it rests upon, developed hand in hand with nature, not in a mythical struggle against it, although that mythology gave capitalist development and its purveyors power. Schumacher, too, saw the need to develop the economy by working within the limits of nature. Nonviolence extended to the environment, and the solution was to keep human development to an appropriate scale.

Schumacher's work focused on the power of the small scale, although he worked from within the UK government to propose alternatives to globalization. After the war, he worked as the chief statistician to the British Control Commission, an agency devoted to rebuilding the German economy. Schumacher was not against industrial development wholesale—it just had to be appropriate to the local context.[56] In England, unlike Burma, apparently it was. Between 1950 and 1970, he was the British National Coal Board's chief economist, a position in which he advocated in favor of reducing dependence on nonrenewable resources like oil—which he saw as key to nonviolent economics. Schumacher's followers saw him as something of a prophet. He noted in 1958 that the cheapest and most plentiful oil supplies were located in some of the world's most politically unstable nations, and he warned against shifting the British energy supply from coal to what he saw as far more limited petroleum resources. In his later work, he (and many of his followers) credited himself with predicting the rise of the Organization of the Petroleum Exporting Countries, or OPEC, and the oil shocks that caused a global recession.

Crucially, coal mining was also a major source of employment in rural Britain, and Schumacher saw the maximization of labor intensity as a strong driver of economic development. Maximizing employment would become a major tenet of his economic philosophy. For Schumacher, technology was essential for development, but only insofar as it did not replace labor—labor-saving technology, he thought, was the main cause of rural unemployment and subsequent urbanization, which decimated traditional subsistence farming. Urbanization, Schumacher thought, led straight to "immiseration" for the masses.[57] Technology disseminated for development programs, then, had to be nonviolent to the web of human social relations as well—it could not disrupt traditional livelihoods.

In 1960, he published an article, "Non-Violent Economics," in David Astor's newspaper, the *Observer*. He exhorted the public to reject the "philosophy of unlimited expansionism" of economics. "A way of life that ever more rapidly depletes the power of earth to sustain it and piles ever more insoluble problems for each succeeding generation can only be called 'violent,'" he wrote. "It is not a way of life that one would like to see exported to countries not yet committed to it."[58] Human needs were limited, he argued, and had to be met through nonviolent means in order to "be secure against a war of annihilation."

∵

Burma had been a transformative experience for Schumacher. Inspired by Gandhi's nonviolence and captivated by Buddhism, he made plans to visit India at the invitation of his friend, Jayaprakash Narayan, whom he had met through David Astor. After a brief visit in early 1961 to deliver an address at the "Paths to Economic Growth" seminar in Poona, he began to think more about India's economic problems and how they could be resolved through nonviolence. In 1962, he took a job as an adviser to Indian prime minister Jawaharlal Nehru on development policy. India had been the front line for many experimental development initiatives both before and after independence in 1947.[59] While Nehru could be described as a high-technologist—at the height of the 1960s, President Kennedy's "Development Decade," modernization was the goal and large-scale industry considered the way to get there—Gandhi's economics advocated change on a much smaller scale.[60] Schumacher continued his study of Gandhian principles of community development through local production and indigenous industry.[61] Although Gandhi was Hindu, Schumacher assimilated what he learned broadly under the aegis of his "Buddhist" economics. Community development, through the Vedic principles of *swaraj* and *swadeshi*, was central to the Indian independence movement. *Swaraj* referred to self-rule and self-restraint—the capacity to govern oneself. This, Gandhi advocated, was achieved by the decentralization of power through a system of participatory democracy at the village level, known as the Panchayat Raj.[62] *Swadeshi*, or self-sufficiency, meant using local means and resources. For Gandhi, a major focus was on using these two principles to promote "villagism," which he saw as a "third way" between the centralized systems of capitalism and socialism.[63]

In India, Schumacher drew particular inspiration from the mastermind of "Gandhian" economics, Joseph Chelladurai Kumarappa, whom he quoted at length in his 1966 paper "Buddhist Economics."[64] Although he rarely received his due credit, Joseph and his brother, Bharatan, were tasked with putting Gandhi's philosophy into actionable practice. The Kumarappas were born into a Christian family in Tamil Nadu. Joseph studied economics at Columbia University in the late 1920s, a few years before Schumacher arrived as a student. His mentor there was Edwin R. A. Seligman, a proponent of progressive taxation. He met Gandhi in 1929 after publishing an article critical of how the British exploited India through tax policies. He wrote *Economy of Permanence* while imprisoned in Jubbulpore Central Jail during the Quit India Movement in 1945.[65] In it, Kumarappa argued that humans see nature as relatively "permanent" given their own relatively short lifetimes, although many of nature's resources are fixed and exhaustible. He outlined five different possible economies given these precepts, the most "violent" and least permanent being a "parasitic economy" in which humans base their existence on exhausting resources, and the most "nonviolent" and most permanent being the "economy of service," which he likened to the relationship between a mother and her young—"it functions neither for its present need nor for its personal future requirement, but projects its activities into the next generation, or generations to come, without looking for any reward."[66] This metaphor was characteristic of the Hindu economists' view of women, who were meant to serve the family and the country with unpaid labor for future benefit. This reproductive labor was essential to the maintenance of the community.

Women did much of the work of health and sanitation, which were central to the village project. Gandhi wrote in 1937:

> An ideal Indian village will be so constructed as to lend itself to perfect sanitation. It will have cottages with sufficient light and ventilation built of a material obtainable within a radius of five miles of it. The cottages will have courtyards enabling householders to plant vegetables for domestic use and to house their cattle. The village lanes and streets will be free of all avoidable dust. It will have wells according to its needs and accessible to all. It will have houses of worship for all; also a common meeting place, a village common for grazing its cattle, a co-operative dairy, primary and secondary schools in which industrial education will be the central fact, and it will have panchayats for settling disputes. It will

> produce its own grains, vegetables and fruit, and its own khadi. This is roughly my idea of a model village.[67]

The health of the population was essential for their ability to work and contribute to the village society. Yet the work of this upkeep—written in the passive voice—rendered women's labor largely invisible. To Gandhi, "the village" did the work, but who worked in the village?

The idealism surrounding villages made its way into economic development programming. Social scientists echoed Gandhi in conceptualizing a "model village," repackaged such that the model could be replicated anywhere, regardless of local context. The ideas drew heavily from the field of "peasant studies," in vogue in the middle of the century.[68] The rural peasants had to be given enough modern science and technology to appease any thoughts of revolt against massive state-building projects, which took on added significance during the Cold War.

After independence, India received large-scale foreign aid investments both through bilateral agreements with the US government's Point Four program, which gave economic and technical aid to developing countries, and through philanthropic outfits. The Etawah model village project in the northern state of Uttar Pradesh has drawn considerable attention from historians. Etawah was designed by the American urban planner Albert Mayer. It contrasted sharply with the "high-modernist" city plans of early twentieth-century urban planners—embodied for James C. Scott by Le Corbusier, who under Nehru's direction modified Mayer's original plans for the Punjab capital of Chandigarh in 1950 to more closely reflect his ideals of monumentalism and "scientific" design, incorporating sunlight, fresh air, plans for electricity, heating, light, and other considerations.[69] Mayer was a regionalist planner, educated as a civil engineer at Columbia University. A friend of the philosopher of technology Lewis Mumford, he was skeptical of large-scale designs that did not take the social and cultural relationships of the city into account. Regionalists, including Henry Wright and Clarence Stein, were concerned with the scale of the neighborhood and its effects on communal welfare.[70] They advocated for small-scale planned communities, both within and outside the United States—other projects included Reston, Virginia, and Stuyvesant Town, New York.

Mayer began the plans for Etawah in 1946, just before independence, as a favor to his friend, Nehru, to whom the project was something of an experiment.[71] He focused on the bottom-up "folk solutions" of the community rather than the top-down advice of outside "experts," which was the standard. Etawah was inspired by Nehru's vision of scientific modernization at the village level. From Lucknow, Mayer consulted with a large number of Indian rural development experts to select the Etawah site, hire staff, and formalize plans. He employed a whole cadre of "village level workers"—"young male Indian agricultural college graduates, each of whom would cover three to four villages on bicycle."[72] Though best known as an American Cold War development scheme, as the historian Nicole Sackley has documented, it remained an entirely Indian undertaking until the geopolitical landscape shifted—in terms of both India's independence and the nascent Cold War. India became a valuable Cold War battleground for the United States, and it was through foreign assistance that the US tried to avoid another communist revolution. In 1952, on Gandhi's birthday, the Indian government announced it would use $50 million from the US Point Four program and a large grant from the Ford Foundation to create more than fifteen thousand additional model villages.[73]

Burma, too, had a program for model villages. The Nathan & Associates economic development plan aimed to build two hundred of them in the Burmese countryside by 1960.[74] "The task of building a New Burma must begin with the village," an excerpt in the public-facing plan read. "We must make our villages pleasant and healthy places to live and work in. We must improve the village dwelling so that it will be more comfortable and durable. We must build schools, health facilities, and community centers in every village. We must construct roads, and bore deep pit latrines and dig wells for sanitation. The new Burmese village must be much better than the old."[75] A "development team" would be on hand in each village to offer advice to villagers on improving their standard of living. Villagers would build their own houses, with the help of five-year loans for durable materials like aluminum and corrugated iron, following blueprints handed to them by the Housing Board.[76] It was to be development financed through debt—both national and personal.

The idea that there could be a functional model village, made to work through technology built on cooperation and local resources, evokes earlier forms of utopianism. The social philosophy of Robert Owen in the early nineteenth century is perhaps the best-known forerunner of cooperative village development. In reaction to widespread poverty and revolts in the English countryside, Owen was commissioned to write a report on the County of Lanark, in which he proposed the creation of cooperative villages. His recommendation was in direct opposition to more laissez-faire approaches to reforming poor laws. In the 1820s, many such communities were built on communitarian principles, including Spa Fields outside of London and New Harmony, Indiana.

The concept of self-help in community development has a longer history, going back to the Antigonish Movement in rural Nova Scotia. The movement began in the 1850s, based around cooperative stores and adult education programs, which culminated in the establishment of the Extension Department at St. Francis Xavier University in 1928. Reverend Dr. Moses Coady was a leader of the movement; he collaborated with the Canadian government to "organize fishermen" by pairing education with spiritual and economic development at the community level. His book, *Masters of Their Own Destiny*, on the movement, was published in 1939.[77] In 1959, St. Francis Xavier University established the Coady International Institute, whose focus is on education for international development professionals from around the world. The Antigonish Movement inspired the thinkers G. K. Chesterton and Hilaire Belloc, who became some of the strongest proponents of an economic theory known as distributism, wherein assets of production are owned widely by the community via cooperatives and mutual associations rather than concentrated in the hands of a few private individuals (capitalism) or the state (socialism). In 1926, Chesterton and Belloc founded the Distributist League, which advocated for decentralized industry in the same way that fellow Catholic Schumacher would make famous half a century later.[78]

The high-modernist ideals of Le Corbusier, Nehru, and other postcolonial leaders in the 1950s and 1960s were offset by an undercurrent of people concerned with community development. Whether through the monikers of *Buddhist economics* or *Gandhian villagism*, they focused on the small scale—a decentralized vision of social order. In 1961, Jane Jacobs published her germinal work *The Death and Life of Great American Cities*. Critical of Le Corbusier's top-down style of high-modernist urbanism, she grounded her analysis in a bottom-up ethnography from the pedestrian city streets.[79] For Jacobs, urban renewal programs and rigid city zoning were an assault on the complex social relations of a city that were visible only from the street-level vantage point. There was a disconnect between the aesthetics of imposed order, as in a rationalist, geometric urban plan, and any form of functional order set up by its inhabitants.

For Fritz Schumacher this gap between the need for community development and the need to respect the social and cultural fabric of a place could be bridged with the right, small-scale technology. On August 14, 1965, he published an article in the *Times* titled "Establishing a New Level of Technology," the first public articulation of the economic theory he had been assembling for over a decade.[80] The paper made the case for "intermediate technologies," designed to be between the "traditional" and the "modern," much in the same way that Gandhi conceived of villagism as a path between capitalism and socialism, and Buddhist theology contemplated the Middle Way of moderation. Intermediate technologies, he proposed, would create jobs in rural communities of developing countries, thus curbing the "immediate cause of much misery" these countries faced—mass urbanization driven by unemployment, a problem he had initially observed in his work at the National Coal Board. However, Schumacher appropriated facets of both Buddhist faith and Hindu economic theories and repackaged them without real consideration for how, concretely, determinations of what was appropriate would be made and by whom.

Two weeks later, in the *Observer*, came "How to Help Them Help Themselves."[81] Considered by many of Schumacher's followers to be the first articulation of his economic philosophy, the article argued that due to misguided priorities, such as modernization, the hundreds of millions of pounds spent on aid to developing countries was hurting rather than helping. Schumacher's argument leaned toward the practical—intermediate technology was that which "would be vastly superior in productivity to their traditional technology (in its present state of decay) while at the same time being vastly cheaper and simpler than the highly sophisticated and enormously capital-intensive technology of the West." Western technology was *too* labor-saving, Schumacher argued, for developing countries facing mass unemployment. He concluded the article by asserting that "'intermediate' technology can help the helpless help themselves."[82]

His idea gained traction within the United Kingdom. In 1968, Oxford University held a conference that aimed to involve British industry in the provision of intermediate technology for the Third World.[83] Followers of intermediate technology as a concept did not shy away from private, for-profit industry but rather saw it as a strategic partner to village-level initiatives. However, a major theme of discussion was the term *intermediate technology* itself—from an engineering perspective, it suggested a technology that was inferior to "high" or "modern" technologies, and it was "criticized for implying a technological fix for development problems, separate from the social and political factors involved."[84] The conference-goers agreed that "appropriate technology" better conveyed the social factors at play in technology selection and eliminated any sort of technological hierarchy, so would likely be better accepted by those who were the intended beneficiaries.[85] In practice, the two words became interchangeable. Virtually every book and paper on the subject of appropriate technology from the 1960s to the 1980s traces the concept back to Schumacher. Even though he stubbornly preferred the term *intermediate technology* in his writings, he used the term *appropriate* frequently when explaining what intermediate technology was, describing it as technology that is "made appropriate for labor-surplus societies."[86] The challenge ahead lay in putting his ideas into practice.

[CHAPTER TWO]

Small Is Beautiful

For Schumacher, the modern world was unsustainable. Technology had fueled economic and environmental disasters that risked the breakdown of world order.[1] He wasn't the only one who thought this way, although for others, breakdown was part of the solution to the world's ills. At the height of the Cold War, Schumacher found a small community of thinkers who put forth alternative views of economic development, including two who would become his close friends and collaborators, Leopold Kohr and Ivan Illich. Each would contribute to Schumacher's thinking about how best to implement technological change.

While Schumacher was at Oxford, the Austrian political scientist and legal scholar Leopold Kohr was finishing his studies at the universities of Innsbruck and Vienna. In 1937, Kohr became a war correspondent in the Spanish Civil War, where he befriended the novelist and playwright George Orwell (who was fighting in a Marxist brigade) and shared an office with fellow journalist (and novelist) Ernest Hemingway.[2] He was deeply impressed by the self-governing small states of Catalonia and Aragon, which inspired his belief in decentralization. By 1938, when the Nazis had annexed Austria, he decided to flee to the United States.[3] Kohr taught economics and political philosophy at Rutgers University from 1943 to 1955, and at the University of Puerto Rico from 1955 to 1973. He came up with the phrase "small is beautiful" that would later make Schumacher famous, and he advocated against the culture of "bigness."[4]

Kohr likened his relationship with Schumacher to a set of Siamese twins, joined at the beards: "Still, even though we had different perspectives, in the end there was a sort of Siamese-twin relationship between Fritz Schumacher and me. The twins [in the article he described] had grown together by their beards, and their beard could of course develop only later in life. So in our youth Fritz and I developed our systems each in our own way and with our own points of view, but later in life it turned out our work had similarities."[5] Where Kohr distinguished himself from Schumacher was in putting his ideas into practice. Fellow Austrian philosopher and social critic Ivan Illich described Kohr as "eminently unassuming" and "radically humble."[6] Kohr maintained that his work remained in the realm of ideas, whereas Schumacher had taken action by launching his own foundation in 1965, the Intermediate Technology Development Group.

This contrast is not quite true, however. In 1966, Kohr and his supporters did indeed attempt to put his ideas into practice. They raised money to support the independence movement of the small Caribbean island of Anguilla, with the hope of creating a model society—a model village on a grander scale—to test his economic and political theories.[7] The attempt was unsuccessful but hints at a key part of Kohr's philosophy—that is, to radically change something that isn't working, it is necessary to break it down before building something new. Illich made this clear in his tribute to Kohr upon his death: "During his lifetime, this teasing leprechaun was recognized by very few as a man ahead of his time. Even today, few have caught up with him; there is still no school of thought that carries on his social morphology. I want to be precise: To place him among the champions of alternative economics would be a posthumous betrayal. Throughout his life, Kohr labored to lay the foundations for an alternative *to* economics; he had no interest in seeking innovative ways to plan the allocation of scarce goods. He identified conditions under which the Good became mired down in things that are scarce. Therefore, he worked to subvert conventional economic wisdom, no matter how advanced."[8]

Kohr was firmly against political and economic unions, preferring instead to break powers down into smaller units. He often referenced his native village, Oberndorf (where, he was fond of recounting, the Christmas carol "Stille Nacht," or "Silent Night," was composed) and the ways in which the inhabitants of neighboring villages around Salzburg would respect the different traditions of each small valley—they were suspect of universal values. Toward the end of World War II, Kohr advocated against the proposal to unite the Allied governments, and similarly against the European Union. He pointed to the breakdown of historical empires to make his argument. "What is the force that changes the course of events?" he asked. "Is it the great leaders—Hannibal, Caesar, Napoleon, Hitler, Stalin, Churchill—who are responsible for the great changes in history? No, they are nothing of the sort! They are the drivers of a defective car whose brakes do not function well so that when it goes downhill, the brakes can make it go faster or slower but cannot stop the descent into the abyss." He was seen by many who also favored decentralization as a prophet for a more federalist Europe. Yet Kohr favored even smaller units than the nation-state.

His 1957 book, *The Breakdown of Nations*, was published only with the help of the English art historian and anarchist Sir Herbert Read.[9] In it, he argued that "bigness" was the fundamental cause of all social misery, and that smaller decentralized social units were the solution. Written in the early years of the Cold War, Kohr lamented the shift from a balance of power among several smaller states to a bipolar world dominated by unified megapowers.[10] He was also critical of the multilateral organizations that had sprung up in the wakes of the two world wars. Of the League of Nations and the United Nations, he said, "None of these glorified vast-scale organizations was ever worth its price, and it makes one shudder to think of the price of an ultimate single World State."[11] Rather, he favored the "reestablishment of small-state sovereignty," which would eliminate many of the highly contested postwar international political debates: "There would no longer be a question of whether disputed Alsace should be united with France or Germany. With neither a France nor a Germany left to claim it, she would be Alsatian. She would be flanked by Baden and Burgundy, themselves then little states with no chance of disputing her existence."[12] So too with Macedonia, Transylvania, and Northern Ireland. With no end to how small a state could be, Kohr argued that dividing states up would also end the constant conflicts over minority rights. Kohr pushed Schumacher to think about implementation at the small scale, which fit naturally with Schumacher's experience in Burma and his championing of village-level interventions.

Ivan Illich was a mutual friend of Kohr and Schumacher. In addition to training as a Catholic priest and historian, Illich was a fierce social critic. Like his friends, his family too had fled the Nazis, and through his extensive publications he became the best known of the three iconoclasts. After taking a job as a priest to a parish of mainly Puerto Rican immigrants in Washington Heights, then one of Manhattan's poorest neighborhoods, he spent several years in Puerto Rico at the invitation of Leopold Kohr; he was appointed vice rector of the Catholic University of Puerto Rico in 1956.[13] In 1961 he founded the Centro Intercultural de Documentación (CIDOC) in Cuernavaca, Mexico—supposedly to serve as a language-learning center for North American missionaries and participants in President Kennedy's Alliance for Progress, a collaboration between the United States and Latin America centered on economic development. In practice, however, CIDOC was a place where Illich attempted to "disorient" the priests and "prospective Peace Corps volunteers" on the harm that US hegemony in the industrial development enterprise wrought on local communities.[14] As Illich got into increasing trouble with the Vatican over his views on birth control and his streak of rebellious community organizing, he sought a space removed enough to allow him more freedom of expression. Cuernavaca was chosen because its bishop, Sergio Méndez Arceo, was known for being open-minded. As was recounted by journalist Francine du Plessix Gray in the *New Yorker* in 1970, "One day, Illich rang the Bishop's doorbell, was ushered into his study, sat down on his couch, and announced, 'I would like to start, under your auspices, a center of de-Yankeefication.'"[15]

Illich invited a curated set of academics to visit Cuernavaca, which became a hub for discussing alternatives to development. Every week, new people would pass through the school as a node between the Global North and Global South. Schumacher would visit on occasion, and he and Illich would invoke Leopold Kohr as if he were at the end of the table.[16] Illich's analyses frequently referred to a principle he termed *specific diseconomy*—that is, when institutions set up to fulfill one purpose actually bring about its opposite. Development efforts, for Illich, frequently resulted in underdevelopment; medicine, in iatrogenic disease.[17] The answer he came to with Schumacher was to scale efforts to fit the purpose and focus on community-driven local needs.

For Illich, though, the corrective was about redistributing authority more than technical knowledge. In medicine and in development, this implied a change in who the primary actors of the enterprise were—not the doctors, but the patients; not the development "experts," but the local people. This created a lot of tension with many institutional models of community-driven development, which focused on distributing the right technologies and technical knowledge (as in later UN-sponsored appropriate technology programs) without ceding authority over the development process.[18] Illich supported what he called "vernacular" technologies, but for him they had to be truly indigenous—conceived of by or with local input, built from local resources, uncomplicated enough that it could be repaired locally, and undamaging to people's social networks.[19] He worried about dependence, complexity, expense, and sustainability.[20]

Jerry Brown, elected California's youngest-ever governor in 1975, was another admirer who visited Cuernavaca. Illich and Brown had first met early into Brown's first term, at the Green Gulch Farm Zen Center in 1976.[21] Dubbed "Governor Moonbeam" for his endorsement of alternative energy and environmentalism, as well as his proposal to invest in satellites for emergency communications, Brown was characteristically cerebral, frugal, and idealistic.[22] He took the threats of climate change and nuclear war seriously in the 1970s, positions that at the time were considered eccentric and attributed to his close association with Illich and the creators of the Doomsday Clock, the *Bulletin of the Atomic Scientists*.[23] Brown had trained to become a Jesuit priest, and he shared with Illich an "indifference to secular values of long life, fame and riches."[24]

During Illich's years in Cuernavaca, he prepared the manuscript that would become *Medical Nemesis*, a treatise that railed against iatrogenic disease. His conversations at CIDOC therefore often centered on medicine and the problems of overmedicalization in Western clinics. In a seminar on demedicalization, the CIDOC group, which then included the man who would become the deputy minister of health under Marxist Chilean president Salvador Allende, discussed applying Schumacherian principles of technology selection to different drugs: their efficacy, side effects, whether they required other supports to take them, the water quality of the region in which they were to be distributed, and their cost. Illich clearly saw modern medicine as a form of dependency—a result of the overreach of industrialization that detached people from the human condition, which included pain, disease, and death. He, and his group at Cuernavaca, helped test Schumacher's ideas with real-world examples and pushed him to consider the implications of technology choice.

∵

Eager to put his new theory into practice, in 1966, Schumacher founded a nonprofit, the Intermediate Technology Development Group (ITDG). Alongside his work at the National Coal Board, he spent his days immersed in a variety of advocacy roles—in addition to chairing the ITDG, he was president of the Soil Association, an old organic farming organization in the United Kingdom; he sponsored the Fourth World Movement, based on his and Kohr's ideas of regionalism and political decentralization; he directed the Scott Bader Company, a Quaker-run global chemical company that restructured in the early 1950s to be worker owned and self-governing; and he continued his studies of Gandhism, nonviolent resistance, and ecological sustainability.[25] In spite of his outreach through the ITDG and the op-ed pages of the *Times*, Schumacher's influence remained largely confined to the United Kingdom until the 1973 publication of his book *Small Is Beautiful: Economics as If People Mattered*.

In the foreword, the historian Theodore Roszak explicitly framed Schumacher as the economist of the "renaissance of organic husbandry, communal households, and do-it-yourself technics whose first faint outlines we can trace through the pages of publications like the *Whole Earth Catalog*."[26] Roszak is widely credited with coining the term *counterculture* in his 1969 book *The Making of a Counter Culture*, to describe the student radicals and hippie movement's rejection of technocracy in the Vietnam War era.[27] Of Schumacher, he said that "it would be no exaggeration to call him the Keynes of postindustrial society, by which I mean (and Schumacher means) a society that has left behind its lethal obsession with those very megasystems of production and distribution which Keynes tried so hard to make manageable."[28]

Roszak framed Schumacher as a libertarian, and that is perhaps the case. By the book's publication in 1973, Schumacher had written that socialism was "of interest solely for its non-economic values and the possibility it creates for the overcoming of the religion of economics. A society ruled primarily by the idolatry of *enrichissez-vous*, which celebrates millionaires as its culture heroes, can gain nothing from socialization that could not also be gained without it."[29] Schumacher's libertarian political economy "distinguishe[d] itself from orthodox socialism and capitalism by insisting that the *scale* of organization must be treated as an independent and primary problem."[30] Socialism and libertarianism seem hard to reconcile, but for Schumacher he drew a distinction between social values, by which he believed socialism would help to free the public mindset from traditional economics, and economic ones, where he aligned more with libertarian principles. While he saw the value in community development and social welfare, he was "hospitable to many forms of free enterprise and private ownership, provided always that the size of the private enterprise is not so large as to divorce ownership from personal involvement."[31] Of course, this concentration of power is the rule in most large bureaucracies, whether public or private.

Given his intellectual lineage, many of Schumacher's arguments in *Small Is Beautiful* are already familiar. He frequently cited Keynes himself, and he saw economics as fundamental to understanding modern society.[32] The key to everything was scale. Societies, factories, and technologies that were too big would ultimately fail. Education was "the greatest resource," echoing the earlier Antigonish Movement and the creation of rural extension schools.[33] The problem with development, meanwhile, was that it was focused primarily on urban areas (to serve the highest number of people the most efficiently) and ignored the root problems in rural areas: mass unemployment and migration.[34] Moreover, he wrote that economic development had to go beyond economics, to education, organization, and what he called "a national consciousness of self-reliance."[35] Development "cannot be 'produced' by skillful grafting operations carried out by foreign technicians or an indigenous élite that has lost contact with the ordinary people."[36] This latter argument is in line with his belief that intermediate technology had to be locally sourced.

He envisioned a shift in international development spending, from mass distribution of food and "material things" to the "gift of useful knowledge," centered on the village.[37] Extending the old saying about giving a man a fish, he wrote: "Teach him to make his own fishing tackle and you have helped him to become not only self-supporting, but also self-reliant and independent. This, then, should be the ever-increasing preoccupation of aid programmes—to make men self-reliant and independent by the generous supply of the appropriate intellectual gifts, gifts of relevant knowledge on the methods of self-help."[38] Education and tools were also less expensive than large-scale aid projects, which Schumacher acknowledged: "This approach, incidentally, also has the advantage of being relatively cheap."[39] Inspired by Schumacher, President Carter attempted to revamp the US Agency for International Development (USAID) programming just a few years later—primarily to save money.[40]

On technology, Schumacher wrote that, on a metaphysical level, "the modern world has been shaped by technology. It tumbles from crisis to crisis; on all sides there are prophecies of disaster and, indeed, visible signs of breakdown"—the last phrase a nod to Kohr.[41] The solution to what Schumacher called this "sick," "inhuman" technology was "technology with a human face"—that which was appropriate and limited. Which was not to say that he thought appropriate technology was appropriate for *everyone*. It was for the poor. Products of "highly sophisticated modern industry" were not amenable to appropriate technology solutions; however, he wrote: "These products . . . are not normally an urgent need for the poor. What the poor need most is simple things—building materials, clothing, household goods, agricultural implements—and a better return for their agricultural products."[42] For Schumacher, the real promise of appropriate technology was in its ability to address global social and economic problems. At the same time, he saw appropriate technology not as "simply a 'going back' in history to methods now outdated," which would be "tantamount to a rejection of science," but rather as forward progress, with science applied in new ways to make technology more accessible and appropriate for "labor-surplus" societies.[43]

In *Small Is Beautiful*, Schumacher synthesized his thinking on everything from "Buddhist economics" to education, unemployment, land use, socialism, and, of course, technology. His arguments were derived mostly from his experiences in Burma and India, but he repackaged his thinking for a broader audience. The book was a sensation. Unlike his newspaper articles, which limited his reach to the British public, *Small Is Beautiful* caught on quickly in the United States. He was accused of being a utopian idealist by some, but the oil shocks of that same year gave his thesis—that the world was consuming limited natural resources too rapidly and that small-scale development was often more appropriate for much of the world—a degree of prescience and relevance that made him a popular hero within the counterculture.[44] On his US speaking tour, organized by his friend Bob Swann and the American Friends Service Committee, he met with California's governor Jerry Brown, Oregon's Governor Tom McCall, several senators, and, as Swann recalled, "even Merrill Lynch & Co."[45] His book was reprinted fourteen times before the end of the 1970s, and it went on to be one of the *Times Literary Supplement*'s "100 Most Influential Books since the Second World War."[46]

Schumacher's reception in the developing world was far more mixed. His ideas about technology fell under especially fierce criticism from expat Marxist economists, who—though not representative of local people—claimed to have more intimate knowledge of the Global South. They were in some ways rivals to Schumacher, although they did not agree on a particular alternative to capitalist economics. Arghiri Emmanuel, a Greek-French economist who lived in the Belgian Congo for decades, became known for his theories of unequal exchange and railed against appropriate technology as "underdeveloped technology; that is to say, one which freezes and perpetuates underdevelopment. This is exactly what should be avoided."[47] Rather, he argued, the only way for the Third World to "catch up" to the developed world was for multinational corporations to transfer the most modern and capital-intensive technology (and therefore, he posited, the technology that maximized the amount produced) to them through the opening of factories in such countries and the hiring of local workers. Multinationals were key to his point, as he believed that factories built by developing countries "by their own means" and with the help of a patchwork of foreign aid grants have historically "broken down": "If Africa has latterly been described as a 'graveyard for machines,' it is to 'direct' projects of this type that it owes the epithet."[48]

By contrast, Samir Amin, an Egyptian-French Marxist economist who lived in Dakar, wrote fervently against technology transfer, arguing that it transferred with it "the underlying capitalist relations of production."[49] For Amin, the high technology that Emmanuel saw as the solution was "excessively costly, not only because of its capital-intensive nature, but because of the wasteful consumption patterns it brings with it, the excessive exploitation of natural resources that it implies, etc. In other words, this technology presupposes imperialism, i.e., the excessive exploitation of labor in the periphery." At the same time, he denigrated appropriate technology, arguing that it was "no solution to borrow the technologies of nineteenth-century Europe."[50]

Schumacher addressed this critique within the pages of *Small Is Beautiful*. He pointed out that the critics were not from the local communities and argued that those who derided appropriate technology as inferior or outdated were not "the voice of those with whom we are here concerned, the poverty-stricken multitudes who lack any real basis of existence, whether in rural or in urban areas, who have neither 'the best' nor 'the second best' but go short of even the most essential means of subsistence. One sometimes wonders how many 'development economists' have any real comprehension of the condition of the poor."[51] In spite of this rebuttal, Schumacher had a hard time winning over intellectuals and policy makers in the developing world, particularly those in the newly independent states of Africa and Asia.

∴

A few weeks after Schumacher departed Rangoon, back in 1955, U Thant chaired the first Asian-African Conference in Bandung, Indonesia. Bandung was significant in that it was one of the first meetings of postcolonial leaders on the two continents, and it reaffirmed their commitment to working together against colonialism and racial segregation, as in apartheid South Africa. In popular memory, the Bandung conference spurred the creation of the Non-Aligned Movement, a bloc of states primarily in what was at the time called the Third World that were not aligned with either the United States or the Soviet Union during the Cold War, and it was formalized in Belgrade in 1961. However, in reality the Bandung-to-Belgrade trajectory is far less clear.[52] The Asian-African movement and the Non-Aligned Movement were, more accurately, rival groups, akin to "a multifront war of position." While the mythology of Bandung often portrays it as the place Josip Broz Tito met Fidel Castro and Kwame Nkrumah, this is the "Paul Revere's ride" version of the story, a "historical distortion," as none of those world leaders was actually present at the conference.[53] Nevertheless, it was a transformative moment for the developing world.

In early September 1961, Thant also represented Burma at the Belgrade conference, where the institutional framework for the Non-Aligned Movement formally began. This is the conference that Bandung is often confused with, as at this one, Yugoslavia's President Tito brought together other postcolonial leaders—notably Prime Minister Jawaharlal Nehru of India, President Sukarno of Indonesia, President Gamal Abdel Nasser of Egypt, and President Kwame Nkrumah of Ghana, all of whom, together with Tito, led the "Initiative of Five." Many other countries, including Cuba under Osvaldo Dorticós and Fidel Castro and Ethiopia under Emperor Haile Selassie, signed on to the cause of Cold War neutrality. The Belgrade conference was organized by Tito and Nasser to explicitly head off a second Bandung conference, which Sukarno had called for, that would have excluded non-

African or Asian countries like Yugoslavia and Cuba.[54] The Non-Aligned Movement gave the Third World a form of collective bargaining power that would be essential to their negotiation of policy priorities, especially science and technology transfer, throughout the 1960s and 1970s.

UN Secretary-General Dag Hammarskjöld was killed in a plane crash en route to cease-fire negotiations in the Congo several weeks later. Thant was elected to replace him. He was the first non-European to hold the top job at either the United Nations or the League of Nations, and the first from a country in the developing world. Interestingly, he wrote in his memoir that "to understand my conception of the role of the Secretary-General, the nature of my religious and cultural background must be understood."[55] For Thant, his Buddhist faith was perhaps as integral to his approach as Schumacher had once advocated it should be for Burma as a whole. Committed to nonviolence, during Thant's tenure as secretary-general, he mediated both the conflict in the Congo and the Cuban Missile Crisis. He was also openly critical of the Vietnam War and frequently was at odds with the American delegation to the United Nations over their refusal to back down.

Thant retired from the United Nations in 1971, after refusing to serve for a third term. Unlike his predecessors, as he stepped down from the post, he was still remarkably on speaking terms with all of the major Cold War power blocs. A military coup in early 1962 had led to the overthrow of U Nu's government and the rise of totalitarian rule in Burma. Thant, like Kohr and Schumacher, was exiled, and he lived out the rest of his life in New York. Until his death from lung cancer in 1974, he continued to be an outspoken critic of apartheid in Rhodesia and South Africa, like David Astor. He spent his final years advocating for a global community—the opposite vision of Leopold Kohr, though meant to achieve the same peaceable ends.

∴

Ernst Schumacher died on September 4, 1977. His memorial, held in Westminster Cathedral on November 30, 1977, was attended by, among other admirers, Governor Jerry Brown, who made the trip to London specially to deliver the eulogy.[56] He had inspired many with his ideas, though it remained to be seen how they could be best put into practice. A *Lancet* editorial, "Community Health Care—Schumacher Style," also served as an obituary for the economist.[57] "He preferred to be thought of as a crank rather than a guru," the author wrote. "A crank, he said, is simply a small mechanism which is hand-operated, but it creates revolution."[58]

The editorial elaborated on the ways that Schumacher's work contributed to human health and well-being, referencing thinkers from

Henry Sigerist to Ivan Illich and even Aristotle where they aligned with Schumacher's ideas. Importantly, they argued that medicine and health were not limited to technological interventions, just as Schumacher's concept of intermediate technology included consideration of the social, cultural, and environmental conditions where the technology was to be used. "Intermediate has now become appropriate technology," the author continued, "the technology includes, not just bits of hardware, but the supplying of knowledge and extra skills to the minds and hands of people."[59] Schumacher, echoing Kohr, had argued for "the freedom of lots and lots of small autonomous units."[60] This would empower people to make decisions about their own local health care systems, and the author thought it was an ideal way to reconceptualize community-based health care. They were not alone. As Schumacher's ideas gained traction, many people—policy makers, technologists, and anticolonial leaders—thought appropriate technology held the promise of freedom and autonomy.

[CHAPTER THREE]

Networking Development

In the early 1960s, a group of engineers began gathering together on their lunch breaks on the Union College campus in Schenectady, New York. Led by Dale B. Fritz, they discussed how they could help address technical problems in the developing world. Fritz was a religious man. Born in Wyoming, he served five years as a marine in the Pacific theater during World War II. Together with his wife, Muriel, he'd traveled since 1953 with his three children to Afghanistan and several other developing nations to give technical advice on local agriculture.[1] He sought like-minded individuals at Union College, and together they formed a voluntary organization, the Volunteers in Technical Assistance, known as VITA.[2] They decided that the best way to disseminate their technical expertise would be to publish a guide—*The Village Technology Handbook*. Published in regular installments with funding from the US Agency for International Development (USAID), the handbook built philosophically on the community development and Gandhian villagism movements that took hold internationally in the 1950s and which had so inspired E. F. Schumacher.[3]

Aimed at development "technicians," the handbook was intended to be "a clearinghouse through which technicians can tell each other of low-cost, locally-developed means of village improvement that they have experienced or observed."[4] The village was key. The foreword to the second edition began: "It is in a world of villages that most of the people live. Surrounding these villages are the farmlands that feed the nations and support their economic development. Thus, any improvement in the villages is of benefit to all." The hefty volume contained over two hundred pages of plans, tips, and blueprints for everything from the construction of community infrastructure—outdoor ovens, privies and latrines, wells, and bamboo pole-frame shelters—to food preservation and soap making.

The handbook attempted to connect development theory with on-the-ground practice. "Those who read and use these handbooks should see that this knowledge reaches down into the village and results in village action," the foreword claimed.[5] Schumacher tried to do something similar, a year later, when he launched his foundation, the Intermediate Technology Development Group, initially based in London. The ITDG, like VITA, wanted to facilitate the exchange of technical information that would enable development practitioners in the Third World to build their own appropriate technologies. The ITDG did this through its own set of publications and eventually an entire publication wing.

In the preinternet era, these sorts of catalogs flourished. Intended to bring the disparate arms of social movements together around a collection of shared resources, the genre facilitated frequent updates and information exchange. Catalogs help illuminate how appropriate technology—as both a concept and a practice—was mobilized into forms of concrete action through key texts published in the 1960s and early 1970s. Their proponents wrote of the "networks" the circulation of the catalogs created between people, which took on new meaning with the advent of the computer. With the aim of increasing "self-sufficiency," the texts were intended to be empowering. Aimed at the individual, the catalog networks nevertheless created a form of community from a distance. Each was written for a specific segment of the broader social justice movement—environmentalists, feminists, and development practitioners. What was notable, however, was the extent to which tools were considered the remedy to social inequalities by these disparate movements, even if the specific tools meant varied greatly. Indeed, the catalog "networks" could be read as their own textual echo chambers, not unlike various corners of the internet today.

At first glance, this embrace of technology would seem to be in contrast with earlier movements of what came to be known as Luddism, which engaged in "machine-wreaking"—particularly the labor-saving factory machinery of the Industrial Revolution. In the eighteenth and early nineteenth centuries, breaking machines was a tried and true tactic of workers' collective bargaining in the period before many trade unions existed.[6] And yet the Luddites, as the historian Eric Hobsbawm pointed out, were not hostile to machines per se. Wreaking machines was simply the most effective way of protesting wage reductions. It put pressure on employers and promoted worker solidarity, as the machines that enabled labor savings and wage reductions came to a standstill.[7] The goal of preserving high rates of employment, particularly in rural areas, was one that E. F. Schumacher shared and one that was integral to his initial conception of appropriate technology. Schumacher's goal was not one of development, then, as it was advanced by foreign aid organizations, but rather one of rural empowerment and maintenance.[8] While this sat well with the privileged (mostly white, middle-class American) followers of the *Whole Earth*

Catalog, and those who were in a position to choose an environmentally conscious, back-to-the-land lifestyle in the United States, it created tensions with those who saw tools as a means of furthering social progress on a broader scale.

Through the networks and catalogs of key organizations, tools—and appropriate technologies in particular—became a central locus of activism. In the United States, this tended toward individual empowerment, whereas in the developing world, it was more focused on benefiting whole communities. Nevertheless, these texts planted the seed among early members of the computing industry, particularly on the West Coast of the United States, of the idea that technology was a transformative means of intervening in the developing world. This would have a pivotal impact later on, toward the turn of the twenty-first century. The irony is that the publication many early Silicon Valley tech leaders credit with starting the West Coast "cyberculture" was heavily inspired by and included material from earlier manuals, like VITA's, aimed at the developing world. The early computing industry used international development networks, and appropriate technology networks in particular, as the model for technology-centric development that it would claim as its own and later feed back to the developing world as innovative.

∵

Dale Fritz and the VITA volunteers were able to secure USAID funding to publish the *Village Technology Handbook* between 1963 and 1988. Focusing on simple, low-tech solutions for health, development, and sanitation, the handbook gave illustrated instructions for how to construct basic latrines, make your own oral rehydration solution, and, in later editions, prevent malaria and bilharziasis by ridding areas of their host organisms.[9] VITA's focus was on technologies that enabled self-sufficiency, to help villagers "to master the resources available to them: to improve their own lives and to bring their villages more fully into the lives of the nations of which they form a basic and important part."[10] In funding this appropriate technology catalog, USAID thus reinforced the idea that appropriate technology for health was that which was low cost, simple, and intended to meet the basic human needs of developing countries' populations.[11]

VITA formed around the same time, and in the same spirit of international volunteerism, as the Peace Corps. In the second issue of the *Peace Corps Volunteer*, published in December 1962, VITA had a full-page spread in which it invited Peace Corps volunteers and other people working abroad to contact them if "confronted with technical problems beyond their ability to solve." An entirely free service, VITA promised to "find someone qualified to study the problem and, if possible, offer a solution." They went on to reprint excerpts of VITA's newsletter in order to give Peace Corps volunteers a better idea of the types of technologies on offer, including plans for building kilns to make bricks, inexpensive production of pesticides, and advice on digging wells safely.[12] While at VITA, Dale Fritz designed his own low-cost, build-it-yourself hand-pump laundry machine and irrigation systems. He would go on to become an agricultural specialist for the Peace Corps in 1970, and his plans for the laundry machine were reprinted in the May–June 1970 issue of the *Peace Corps Volunteer*.[13]

In February 1968, Congressman Daniel Button of New York lauded VITA on the House floor for its provision of "effective, economical foreign aid to the emerging nations of the world." He had visited with Dale Fritz and other "officers and friends" of VITA earlier in the month at Union College, and he was clearly impressed. He cited the group as an "inspiring example of responsible and constructive cooperation between the government and the private sector," which President Johnson had echoed in his message to Congress on foreign aid. As Button pointed out, President Johnson had also said in the address: "Foreign aid must be much more than government aid. Private enterprise has a critical role." Aside from the USAID grant, VITA received 70 percent of its funding from the private sector and was an effort built on volunteer labor. Button expounded on the "multiplying effect" VITA had—"VITA multiplies the AID dollar by at least a factor of 10. AID's support makes it possible to extend this voluntary service to many more people . . . The solutions, once demonstrated in one setting, can be multiplied by local exposure and through VITA's publication program."[14] At the Union College campus, he praised the group for providing the government with "the most *meaningful* service in the most economical manner; the kind of approach we would hope to make in *all our* dealings" (emphasis his). Further, he let the group know that he'd heard that VITA's services would soon replace the Technical Inquiry Service that USAID had maintained for many years—an early example of the agency contracting out vital capacity to the private sector.[15]

The organization continued to grow expansively. By 1968 it had several thousand members across every state in the United States. It formalized a contract with the Peace Corps wherein the Peace Corps would subsidize half the cost of handling one thousand inquiries that Corps volunteers submitted to VITA volunteers. Button called this "probably the most service Peace Corps has ever received for $20,000, plus having the additional advantage of providing this help without affecting our adverse balance of payments." "All of VITA's work is foreign aid with no gold outflow," he added.[16]

The year 1968 was otherwise a tough one for the foreign aid budget. After the Tet Offensive in late January of that year, the tide of public opinion had turned against the Vietnam War as Americans suffered heavy casualties. Involvement in foreign aid was seen as wasteful, particularly as disillusionment with the heavy spending on the modernization projects of the 1950s and early 1960s set in. Between the 1968 and 1969 fiscal years, the USAID budget was slashed by 26 percent in inflation-adjusted dollars.[17] The heavy spending on the war had created balance-of-payments problems for the US government more broadly, so officials were keen to keep spending home, where possible. VITA and other US-based contractors were, therefore, a very appealing solution, as they were inexpensive and kept government dollars in the US economy.

VITA also quickly won the admiration of Schumacher's own foundation, the ITDG, which, too, saw a major role for the private sector in the provision of appropriate technologies. An early ITDG publication straightforwardly subtitled a piece: "Wanted: A British VITA."[18] The opening tribute focused on the networking aspect of the VITA handbook: "In Guatemala there is a bridge built by local people from local materials. It was designed by an engineer from Maine who never went near the place. A dozen bridges in Sierra Leone, also built by local people with local materials, owe their existence to a University of Texas professor of civil engineering and some of his students. The bridges were built by correspondence—the designers never left home."[19] The ITDG's fascination with VITA's ability to correspond and share plans via an international network meant that it placed a rather higher value on the American designers of the infrastructure projects than on the workers who actual built the bridges and realized the plans in Guatemala and Sierra Leone.[20]

The *Village Technology Handbook* was a way of sharing plans, ideas, and technical expertise, with many diagrams and blueprints drawn to scale. The instructions were meant to be simple and easily implemented by villagers with basic tools and local materials. VITA ran its Inquiry Service, in which people from around the world could write in asking for advice or a solution for a specific development need. However, the VITA volunteers soon found that they did not have ready-made blueprints or designs for all situations. They also realized that they needed a way to test the plans that their Publications Department had been collecting for dissemination in the handbook.

In 1968 VITA launched a Village Technology Center in Schenectady, where it prototyped equipment and tested designs that had been sent in. This, too, was a model village—but wholly new. Unlike the scientific designs of Etawah or the economic and engineering plans of Nathan & Associates in Burma, the Village Technology Center did not build or seek to improve upon an extant traditional village. It was a lab. Of course, it also published an accompanying catalog.[21]

The *Village Technology Center Catalog* described the technologies that VITA volunteers had tested and identified, and it offered the option of supplying plans for the item, renting it out, or selling it ready-made. Examples include concrete-block molds made of metal, the "CINVA-RAM block press" that made building blocks from dirt, and a "beehive building."[22] In addition, VITA began running training programs out of the technology center for contractors, students, and "persons going overseas in general," and consulting on how to set up similar centers in developing countries.[23] Also in 1968, Dale Fritz helped establish a center at the International Institute for Rural Reconstruction training site in Silong, Philippines, about twenty miles south of Manila. It was to be a permanent training site, teaching people "better farming methods, better health practices, literacy, and self-government," and training leaders for other institute programs in Colombia, Thailand, and Ceylon, as well as the YMCA, Peace Corps, and missionary groups. The center featured "18 full-scale models of devices useful in rural areas—including a water pump made from bamboo and a bicycle-powered wood lathe."[24] Ever the pragmatist, Fritz's do-it-yourself devices were also model appropriate technologies: built from locally sourced materials, by the local community, and efficient but not labor-saving to the point of putting anyone out of a job.

Subtitled "Tools for Development," VITA's catalog was featured in another text concerned with access to tools—the *Whole Earth Catalog*. In the first issue from fall 1968, sandwiched between information on geodesic domes and "Indian tipis," a full page was dedicated to VITA, titled "Village Technology." *Whole Earth* author Stewart Brand wrote: "VITA is the only source of specific practical information on small-group technology that we've found. But what a source." He goes on to say that the *Village Technology Handbook*, though intended for use overseas, "is ideal for rural intentional communities." The *Village Technology Catalog*, meanwhile, was full of "funky tools."[25] The rest of the page was filled with excerpts from the catalog, giving readers an idea of the types of technologies available. Many of the same excerpts were reprinted by Schumacher's own foundation, the ITDG, when it was looking for an organizational model to put appropriate technology theory into practice.

∴

Schumacher had founded the ITDG, a nonprofit foundation based in London, in 1966 to advance his ideas of appropriate technology in practice.[26]

The group's first-line strategy for doing this was to start a bulletin, mailed out to members from headquarters in Covent Garden. Membership cost one British pound per year for an individual, or five pounds per year for a corporation or firm.

The first issue, in September 1966, contained just three articles: a reprint of a speech Schumacher had delivered to the United Kingdom's Africa Bureau, instructions on how to build rainwater catchment tanks, and a tract by the missionary Reverend Lindsay Robertson in Malawi that explained why education about the utility of intermediate technology was needed for both UK and African elites. In this latter, Robertson used the example of ox carts and a description of how his mission constructed them from a mix of local materials and "that little bit of material and know-how from our civilization—namely, a few bolts, a draw-bar, of course, the axle, the wreck of an old car from our civilization!"—this to argue that intermediate technology was underappreciated by African and British elites alike because it was not "modern."[27] He explained that his mission produced the carts for about twenty-nine British pounds and sold them for thirty to thirty-five, depending on the size and the state of the tires. Given that importing a cart cost upward of sixty pounds, Robertson felt that they were helping both the local "underemployed" society and the mission, which made a small profit. The *ITDG Bulletin* also noted that it was making the plans for the ox carts available for any correspondents who wanted them.

The second issue, published in March 1968, was subtitled "Tools for Progress 1967/68." It included an update on the rainwater catchment tanks of the first issue—a pilot project was underway in Botswana—and news of ITDG's expanded activities.[28] The main article promoted the ITDG's first major publication, launched in December 1967, also titled *Tools for Progress*. Billed as "the first-ever guide to low-cost tools and equipment available in the UK," with most items costing less than a hundred pounds, *Tools for Progress* was intended to be a comprehensive catalog of off-the-shelf solutions for agriculture, small-scale industry, health, and education. The catalog came in at 190 pages and was indexed in four languages (English, French, Spanish, and Arabic). The ITDG shipped three thousand copies of the catalog overseas, with the bulk of copies bought by the UK government's own Ministry of Overseas Development for distribution to its field experts.[29]

Tools for Progress was, of course, not the first such field guide available. VITA's *Village Technology Handbook* had been published and distributed overseas through American development programs since 1963. However, the two publications differed in the extent to which development practitioners and local beneficiaries were supposed to construct and assemble the technologies themselves. While VITA's guide was mostly one of information and open-source blueprints, ITDG's was more of a commercial catalog, with the contact information of British firms where the technologies could be purchased. The ITDG thus advanced the idea that the private sector played a major role in furnishing appropriate technology.

In the second issue of the bulletin, the ITDG also described its efforts to build bridges with like-minded organizations, including VITA, and to foster local instantiations of appropriate technology groups, as had been formed by that point in India, Colombia, and Peru. It hosted a conference on intermediate technology at St. Cross College, Oxford, on January 12 and 13, 1968, to which it invited representatives of VITA (who had given financial support to ITDG's Indian subgroup), the British Ministry of Overseas Development, the UN Economic Commission for Africa, and representatives of British industry and the British National Export Council. Unlike their American counterparts at VITA, the ITDG saw a large role for private industry in the provision of intermediate technology—it was not a do-it-yourself alternative to patented technologies, but a new, if constrained, way of marketing them.[30] *Tools for Progress* was thus a "buyer's guide" for rural areas, aimed at reversing the "collapse of rural life" that Schumacher saw as the root of mass emigration, unemployment, and undernourishment.[31] As with his work in Burma, Schumacher embraced the possibility of working with the private sector on appropriate technology solutions.

By the third issue of the *ITDG Bulletin*, in August 1968, there were reports on the formation of a fourth overseas group that had been fostered by the organization, run out of the University of Science and Technology in Kumasi, Ghana.[32] Known as the Kumasi Technology Group, the new nonprofit led by Dr. P. Dziwoonoh would enable members "with specialized technical knowledge in science, agriculture, pharmacy, architecture, and the useful arts" to provide advice and assistance to indigenous Ghanaian entrepreneurs in establishing small-scale industrial outfits "without heavy capital."[33] The secretary of the group, a Dr. J. Latham, said that the group members believed that "a completely unwestern approach" was needed to make progress in Ghana's industrial development.[34] The Kumasi Technology Group's aim was to manufacture as much as possible locally to employ local people and replace expensive imports.

Two years later, the ITDG helped to advance a proposal by Professor S. Sey, dean of the School of Agriculture at the University of Science and Technology in Kumasi and member of the Kumasi Technology Group, to get funding from the UN Economic Commission for Africa (UNECA) and the UN Industrial Development Organization (UNIDO) to transform the group into an official technology consultancy center at the university. Until that point, as Schumacher's friend and ITDG affiliate George McRobie wrote in the proposal, the group had been entirely voluntary and regarded by the university as extramural.[35] At the UNECA meeting in 1970, Professor Sey presented the proposal in person. Advocating for greater recognition by the university and funding from UNECA, Sey argued that the university should recognize faculty contributions "as useful and essential" advisory on "problems which may not be immediately of academic interest but nonetheless is of vital importance to the economy of the country."[36] The bid for support was eventually successful. ITDG leveraged the Kumasi Group to gain UNIDO funding via UNECA not only for its center but for ITDG-led projects in Togo and Nigeria as well.[37] The Technology Consultancy Centre, officially founded in 1972, is still a vibrant part of the Kwame Nkrumah University of Science and Technology today.[38]

The ITDG was also successful in winning grant-based funding from the Ford Foundation and the British Ministry of Overseas Development, in addition to its own fundraising. Throughout the 1970s the ITDG continued reporting on and funding small-scale technology projects, such as the work of instrument engineer S. W. Eaves at the Ahmadu Bello Hospital in Nigeria. Claiming his work as evidence that "nothing succeeds like self-help," the *ITDG Bulletin* described his locally made hospital equipment—from wheelchairs to surgical vacuum pumps—made from repurposed materials like sheet metal, bicycle wheels, and car brakes.[39] The nascent publishing wing, ITDG Publications, put out low-cost guides for everything from community housing to smallholder agriculture to how to make a cookstove.[40] By 1980, the ITDG had taken on the Prince of Wales as a patron and was expanding operations overseas with field offices and work on the ground (for more on this, see chapter 4). The ITDG's efforts, along with appropriate technology programs put in place by VITA, the US government, and a host of other nongovernmental organizations, were chronicled in a 1981 volume entitled *Small Is Possible*.[41] Written by Schumacher's ITDG colleague George McRobie and with a foreword by Schumacher's widow, Verena, it was billed as "the third in the late E. F. Schumacher's planned trilogy begun in *Small Is Beautiful* and *A Guide for the Perplexed*." The book served as a showcase to the variety of ways in which appropriate technology was being interpreted and used in the service of international development.

∴

Although appropriate technology was conceptually intended for the developing world, as the movement was taking off in the 1960s, one of the key figures of what would become the countercultural movement in the United States was keen to explore how it could be applied within Western society. In February 1966, twenty-eight-year old Stewart Brand, on a mild LSD-induced high, sat on a roof in San Francisco and looked up at the city's skyline. "As I stared at the city's high-rises, I realized they were not really parallel, but diverged slightly at the top because of the curve of the earth," he said. He recalled imagining going up into orbit to see the curve of the earth more defined, and "soon realized that the sight of the entire planet, seen at once, would be quite dramatic and would make a point that Buckminster Fuller was always ranting about: that people act as if the earth is flat, when in reality it is spherical and extremely finite, and until we learn to treat it as a finite thing, we will never get civilization right."[42] He thought that a color photograph of the whole earth could be the solution—something that would shift people's perspectives: "There it would be for all to see, the earth complete, tiny, adrift, and no one would ever perceive things the same way."[43] He fashioned buttons that read "Why haven't we seen a photograph of the whole Earth yet?" to lobby the National Aeronautics and Space Agency to release a photo from space. He sold them at the University of California at Berkeley for twenty-five cents each, and soon branched out to Stanford, Columbia, Harvard, and the Massachusetts Institute of Technology. He sent buttons "to scientists, secretaries of state, senators, people in the Soviet Union, UN officials, and famous thinkers like Marshall McLuhan and, of course, Buckminster Fuller." The latter wrote back, "Well, you can only see about half the earth at any given time."[44]

Brand recalled that the photo was eventually taken by the "homesick" Apollo 8 astronauts in 1968; however, the image that adorned the front cover of the first edition of Brand's *Whole Earth Catalog*, published in the fall of 1968, was actually taken by an ATS-3 satellite on November 10, 1967.[45] For Brand, putting together a catalog was a way of connecting like-minded people across space and political movements—he was fascinated by networks.[46] In imagining the service it would deliver, he said he was directly inspired by the L.L. Bean catalog.[47] The first edition of *Whole Earth*, subtitled "Access to Tools," also drew not just inspiration but photos, full quotes, and listings from VITA's *Village Technology Handbook*. Like VITA's publications, *Whole Earth* carried listings for off-the-grid products, although they were often finished products for purchase (as in the ITDG catalog) rather than blueprints. It also featured several articles each by Stewart Brand and Buckminster Fuller. Trained as a biologist at Stanford, Brand was strongly influenced by the engineer and philosopher Fuller's ideas about efficiencies of whole systems and "synergetics." Early editions of the catalog reflected this viewpoint. However, after the publication of *Small Is Beautiful* in 1973,

Brand began reading the work of E. F. Schumacher, whose ideas of "Buddhist economics" and more environmental, moral motivations softened the more coldly calculated principles of engineering efficiency.[48] The catalog was systems oriented, reflecting the intellectual framework and social practice of cybernetics, an "alternative form of communal organization" through networking.[49]

By the September 1974 edition, *Whole Earth Epilog*, there was an entire section of the book dedicated to "soft technology." The ITDG featured heavily in the discussion of soft technology research communities, and many of the ITDG publications were found for sale on *Whole Earth*'s pages. As with VITA, Brand wrote that ITDG "furnishes a marvelous proliferation of goodies," which, though aimed at the developing world, were equally suitable in his view to a tech-savvy return to rural America.[50] *Small Is Beautiful*, the first of Schumacher's writings to make it big in the United States, was also for sale in the *Epilog*, featured right next to Ivan Illich's *Tools for Conviviality*.[51] Although Brand repackaged Schumacher's message for his American audience, as with VITA before, the baseline reality for subscribers to the *Whole Earth Catalog* was fundamentally different from that of the people living in Burma and India who had inspired it.

Stewart Brand and the *Whole Earth Catalog* are often credited with incubating the first generation of tech entrepreneurs in Silicon Valley.[52] The networks of Brand's "cybernetic counterculture," according to the cultural historian Fred Turner, spawned early visions of computer networking and the "cyberculture" that resulted. By networking people and ideas, you could network machines. Through the catalog, Turner argued, Brand supplied the space for the counterculture to align with capital, technology, and the state—computers were no longer considered solely tools of the Cold War military-industrial complex but a means of achieving digital community. In focusing solely on the United States, however, Turner overlooked the fact that Brand borrowed extensively from earlier catalog-based networks aimed at international development. Likewise, the historian Andrew Kirk placed Brand and the *Whole Earth Catalog* at the center of the countercultural environmental movement of the late 1960s and 1970s.[53] For him, *Whole Earth* not only defined but also was the main purveyor of appropriate technology for pragmatic environmentalists. While this is perhaps true for a certain demographic—American, male, middle class, and white—the narrow frame misses the far broader reach of appropriate technology.

Inevitably, Brand's own preoccupations take predominance in the pages of *Whole Earth*. Pages on cybernetics, computers, hunting, guns, knives, and other gadgetry abound. In the *Epilog*, there are twice as many pages devoted to sex—with most books personally blurbed by Brand—as there are to health. Brand remarks on their inclusion (and the fact that the sex and birth pages feature what he calls "obscene" photos and graphics of naked women) on the first, orientation page of book, asking readers not to take offense. "If it's not your taste, please overlook it," he wrote, "then no one need complain, and no one need apologize."[54] Only one page is devoted to "women" (the same number as devoted to the game go).[55]

Brand famously opened the first edition of the catalog in 1968 with the declaration "We *are* as gods, and we might as well get used to it. So far, remotely done power and glory—as via government, big business, formal education, church—has succeeded to the point where gross defects obscure actual gains. In response to this dilemma and to these gains a realm of intimate, personal power is developing—power of the individual to conduct his own education, find his own inspiration, shape his own environment, and share his adventure with whoever is interested. Tools that aid this process are sought and promoted by the *Whole Earth Catalog*."[56] In a later essay on the Whole Earth website, he admitted that he stole the opening line, mostly, from British anthropologist Edmund Leach, whose opening line to his 1968 book *A Runaway World?* was "Men have become like gods."[57] It echoed Schumacher's critique of *Homo sapiens*, scientific man, although it's not clear Leach was aware of it. An expert on Burma and former member of Bronislaw Malinowski's "famous seminar," Leach was regarded as an iconoclast by many of his contemporaries. In *A Runaway World?* he inquired as to the relationships between humans and the environment in the face of massive population growth, and the promises and perils of technological advancement. Although men were like gods, Leach contended, this very isolation from the natural world created fear, which resulted in further division and self-withdrawal.[58]

It is somewhat ironic, then, that Brand's catalog was built to enable this withdrawal into idealistic, intentional communities—at least, for those who could afford the tools. To Brand, this individualism was the preferred way to address overarching social problems like climate change.[59] Technology could help overcome limits, whether ecological or economic. He wrote, "At a time when the New Left was calling for grass-roots political (i.e., referred) power, Whole Earth eschewed politics and pushed grass-roots direct power tools and skills. At a time when New Age hippies were deploring the intellectual world of arid abstractions, Whole Earth pushed science, intellectual endeavor, and new technology as well as old."[60] Individualism and personal liberty had long motivated Brand's actions—animating his fears of the Soviets as an adolescent.[61] He was not interested in broad-based political activism.

The most sustained criticism of the *Whole Earth Catalog* and its legacy cult following in Silicon Valley is derived precisely from this point. It was one thing to tell those with the means to *choose* to go "back to the land" to "Stay Hungry. Stay Foolish." (as the back cover of the *Whole Earth Epilog* advised in 1974). It was another to hold that up as a model in a world marked with widespread inequality, poverty, and very real hunger. Held up as an early inspiration for libertarianism for many billionaire tech entrepreneurs today, Brand rejects the continued relevance of *Whole Earth*'s politics. In October 2018, more than one hundred *Whole Earth* network alums gathered at the San Francisco Art Institute to celebrate the catalog's fiftieth anniversary. That evening, as journalist Anna Wiener reported in the *New Yorker*, Brand addressed the crowd, detaching himself from the movement: "There's a kind of reframing, I hope, going on today, in relation to 'Whole Earth,' and it's getting over the sense that somehow the 'Whole Earth' is Stewart Brand," he said.[62] In a conversation with Wiener after the event, Brand described *Whole Earth* as "well and truly obsolete and extinct." He seemed dismayed that he was facing criticism and blame for the darker underbelly of the modern techno-libertarian movement, including rampant sexism and behemoth tech monopolies. "The people who are using my name as a source of good or ill things going on in cyberspace, most of them don't know me at all," he told Wiener. "They're just using a shorthand. You know, magical realism: Borges."[63] This octogenarian Stewart Brand is still known for his rugged practicality: dressed characteristically in tan Levi's and a flannel shirt, he carries his glasses in a tool case hanging by a carabiner from his belt, his Moleskine notebook and Swiss Card multitool in a black leather satchel, and he attends CrossFit classes regularly. His politics, though, he says have changed. Reflecting back on his political development, he credited his work as an adviser to Governor Jerry Brown of California in the 1970s with shifting his opinion away from libertarianism.[64] Since that time, he claimed he has had a renewed respect for social institutions and the government in particular. The individualism represented in the pages of *Whole Earth* he thus attributed to youthful hubris: "We didn't know what government did. The whole government apparatus is quite wonderful, and quite crucial. [It] makes me frantic, that it's being taken away."[65]

That Jerry Brown sat at the inflection point of Brand's political trajectory is not surprising. The governor shared Brand's penchant for environmentalism and technological solutions, and during his time in office he met with and admired Schumacher. He had visited Cuernavaca and participated in Illich's late night salons. Yet for the most part, he exercised his ideological convictions through the mechanisms of state government. If anyone could win Brand over to the importance of the state, Brown could.

And though Brand now identifies as "postlibertarian," his lack of proactive support for the broader social movements of the 1970s has been a more difficult legacy to shake off.

∵

Brand's *Whole Earth Catalog* focused on providing access to tools for individual empowerment. Tools were supposed to be neutral. He wasn't interested in contributing to broader political movements of the day—anticolonialism, feminism, or civil rights. As is often typical of middle-class American white men, Brand did not see the importance of directly challenging the status quo to rectify power imbalances. The *Whole Earth Catalog* inspired a network of back-to-the-land computer enthusiasts who, unsurprisingly given its politics, were predominantly men.

In 1973, Susan Rennie and Kirsten Grimstad published the first edition of the *New Woman's Survival Catalog*. A feminist take on the *Whole Earth Catalog*, Rennie and Grimstad began the catalog to "document activities which, unlike women's businesses and enterprises that have existed all along, are aimed explicitly at the development of an alternative women's culture." They stood against institutional structures, but "unlike the male hip counter-culture, represent[ed] an active attempt to reshape culture through changing values and consciousness."[66] In their introduction to the catalog, they tie the growing number of women "forming their own law firms and legal clinics, establishing their own business companies, running their own printing presses, publishing their own magazines and newspapers, starting their own credit unions, banks, anti-rape squads, art galleries, and schools, hospitals, non-sexist playgroups and child care centers, bands, theater groups, restaurants, literary magazines, and scholarly journals" to changing views and potential for women to control their own reproductive capacity. Confinement to the roles of wife and mother, they argued, created "massive discontent affecting a wide spectrum of American women—even those who would most vehemently and indignantly deny any affinity with feminism."[67]

For Rennie and Grimstad, the book was "a tool for women whose rising expectations are running into the wall of patriarchal privilege." However, this book-as-tool was not just to develop personal skills and intentional communities, as Brand fashioned his tools. The services that Rennie and Grimstad cataloged were what they called "survival tools," which were becoming more and more necessary in the early 1970s. "Rape rates are going up; rape arrests are going down," they wrote. "The number of women who must enter the labor market is going up; women's proportionate earnings are going down."[68] The tools and services on offer in their catalog would help cure women of their "false consciousness" and dependency on men, which was the result not of any real helplessness but of inculturation. Unlike the tools on offer from *Whole Earth*, the listings were rarely for physical implements. Rather, they were tools for spreading knowledge and a movement. In that way, it was more akin to VITA's *Village Technology Handbook*, although there's no evidence that Rennie and Grimstad were aware of that publication.

With listings for everything from women's radio programs and print publications to "self-health" guides and self-defense classes to legal and job-seeking resources, the catalog aimed to build women's self-sufficiency. The goal was not unlike that of the newly independent states that aimed to reduce their dependency on foreign aid and colonial governments through technology. Anti-oppression movements are intrinsically linked. And yet it wasn't until the second edition of the catalog, in 1975, that author-compilers Rennie and Grimstad made any reference to the Third World.[69] Writing primarily about "Third World women living in the United States," a catchall term they use for women of color, they write that women "who confront sexism *and* racism on a daily basis, must, by necessity, have a significantly different analysis of their oppression than do white women. It is important for us all to understand why, in a racist society, the *forms* of feminism chosen by these two groups (minority and white women) will continue to differ."[70] Citing an article by Angela Davis in a 1971 edition of the journal the *Black Scholar*, Rennie and Grimstad go on to critique the Moynihan Report—a report on black poverty in the United States written by the sociologist Daniel Patrick Moynihan for the Johnson administration's War on Poverty in 1965—for its analysis connecting black matriarchy with juvenile delinquency and the sexism on the part of black male "revolutionaries" that it inspired.[71] They also point out that, "despite their ample reason for distrusting whites," black women did support the feminist movement—at a higher rate, by a 1972 poll, than white women did. So-called Third World women, both within the United States and in the developing world, were cognizant both of the racism within segments of the feminist movement and of the sexism within segments of the anti-imperialist liberation movements. In an effort to generate more widespread understanding of this intersectionality avant la lettre, Rennie and Grimstad provided a reading list for their readers to take up.

The second edition also paid far greater attention to the experiences of lesbian women and the ways in which their politics and movement did and did not intersect with the contemporary feminist movement. With all new materials, Rennie and Grimstad framed the second edition, *The*

New Woman's Survival Sourcebook, as going "far beyond cataloging the resources, enterprises, and services which establish the women's movement as an irreversible social phenomenon. The sourcebook is an inventory of the *ideas* of feminism. Specifically, it is the documentation of the incredibly exciting and significant emergence of female consciousness."[72] While there are listings for resources developed by or for lesbians and women of color throughout the sourcebook, they devote one section to the ten-paragraph manifesto *The Woman-Identified Woman* written in 1970 by the collective known as the Radicalesbians.[73] It argued that the woman's movement could not be successful with the exclusion of lesbian voices, because lesbians—by defying patriarchal structures in which women identified via their relationships to men—centered women's relationships to other women. The creation of a new sense of self, through women's collective consciousness and the formation of a women's culture, was to the Radicalesbians the key to liberation.[74]

Sections on self-health dealt overwhelmingly with women's reproductive health.[75] Encouraging women to understand and "love" their bodies, listings both conveyed information for feminist health centers and services (including a directory of feminist abortion providers) and instructed women on how to care for themselves. Excerpts from books and articles decried legal rulings against feminist health practitioners for practicing medicine without a license, described where to find one's clitoris and how to masturbate effectively, and diagrammed where different forms of birth control fit within men's and women's anatomy. Much like *Our Bodies, Ourselves,* the 1970 book written and published by the Boston Women's Health Collective, the catalog sought to empower women to not only control their own reproduction and sexuality but also seek out alternatives to conventional medicine.[76] Several listings questioned the safety of the pill, diaphragm jelly, and douches. Others advised women on how to do vaginal self-exams with a speculum.[77] The book and catalog were technologies used for self-sufficiency and building community—a Western, feminist appropriate technology.

The *New Women's Survival Catalog* encouraged women to educate themselves, to learn skills they could use to provide for themselves and other women, and to seize control of tools that would enable them to lead independent lives. The control of technology and machines, which under patriarchal systems had been granted almost exclusively to men, was perhaps more radical than it seemed. Indeed, some analyses of the Luddites have argued that it was in part the fact that "new machinery made it possible for employers to use cheaper unskilled workers—women, and even children—in the place of highly paid artisans" that encouraged machine breaking as a tactic.[78] Even in later instances when women exerted significant control over technology, as in the early computing industry in Britain during World War II, that power was usually short-lived.[79]

In a 2018 interview with the journalist Meg Miller, Kirsten Grimstad and Susan Rennie explained that selling the book to the publisher as the "woman's *Whole Earth Catalog*" was the major hook.[80] They'd begun the project as a feminist bibliography for the newly opened Barnard College Women's Center when they were both graduate students at Columbia. And while they thought it had to be activist, otherwise it would not be truly feminist, Grimstad recalled that "the brilliance of the *Whole Earth Catalog*, for our purposes, was the networking element. That's what our book strived to do: to create some sort of nationwide network of feminist alternative culture."[81] For Rennie and Grimstad, *Whole Earth* was the publication that had innovated the catalog network—Brand's inspiration from VITA wasn't discernible. When the *New Woman's Survival Catalog* was featured in *Whole Earth*, however, the editorial commentary (written by Carole Levine, not Brand himself) celebrated the concept but also lobbed criticism at Rennie and Grimstad for using the *Whole Earth* model: "I would have hoped the ladies could have turned out something more than another in a long line of *Last Whole Earth Catalog* imitations, down to the graphics and print type (albeit better than most of the imitations flooding the market!)," Levine wrote.[82]

Appropriate technologies are locally sourced and locally made by the community they're intended to serve. As suggested by the subtitle, "A Woman-Made Book," Rennie and Grimstad paid careful attention to the catalog's production. They "aspired to a handmade quality," since "so much of what women were doing throughout the country was do-it-yourself." Along with two interns at Barnard, they wrote, edited, typeset, and screen printed the book themselves in about five months—three months of research and two months of production in a small apartment between Harlem and the Upper West Side. "The bathroom was turned into a darkroom. We rented a machine from IBM that was called a Composer, and a woman named Mark St. Giles typeset the magazine . . . What we put together for the publisher was camera-ready copy. At the end we took it all in a big box to the publishers in a taxi. Then we went to the Russian Tea Room and fell asleep at dinner, we were so exhausted," Grimstad remembered. A rather posh New York institution, the Russian Tea Room hints at Rennie and Grimstad's relative privilege. They went on to found the feminist magazine *Chrysalis*, which they ran out of the Women's Center in Los Angeles.

While the *Whole Earth Catalog* retains a strong counterculture following in the present, Miller asked Rennie and Grimstad why they believed the *New Woman's Survival Catalog* had not had a similar legacy. Susan

Rennie pinned the decline of feminist groups and the backlash against the movement to the election of Ronald Reagan.[83] Funding cuts to federal community development grants destabilized groups like the Women's Center, while "feminism was being called 'the f-word,'" Grimstad said. The election of Ronald Reagan in 1980 and the ascendance of what would come to be called neoliberal economics also caught blame for the decline in funding for international health and development programs and for policies that supported business growth at the expense of the environment.[84] These movements undoubtedly were affected by the shift in presidential administrations and macropolitical leanings; however, the rifts created by advancing individualism and the growth of the private sector—at the expense of community and public goods—had begun in the late 1960s.[85]

Schumacher had warned in *Small Is Beautiful* that large-scale technological systems, particularly when married to the state, could take on a life and scale of their own—ruling human life rather than being ruled by them. Individualism could be seen as a countervailing force—and networks a way of mobilizing against oppressive structures. Feminism, environmentalism, and remedying world poverty are not causes whose interests typically align with untethered global capitalism. Text-based networks, centered on the circulation of common handbooks, catalogs, and ideas, helped rally a critical mass of support for these causes.

∵

Schumacher saw appropriate technology as a way out of cycles of dependency for the world's poor, so long as they were content to remain small-scale farmers and artisans. As implemented by VITA and the ITDG, however, most seekers of appropriate technology were still dependent on engineers and expertise from the United States or United Kingdom. While its theoretical definition remained stable, questions remained about what, exactly, appropriate technology was in practice. Putting the idea into practice required a degree of interpretation, even within the networked communities aimed at international development. The catalogs offer a glimpse into how diverse these interpretations were. But significant controversy existed over who got to make these determinations—who decided what an appropriate technology was? And for whom? These debates emerged, as we'll see in the next two chapters, most forcefully in transnational and multilateral forums over the course of the mid- to late 1970s.

[CHAPTER FOUR]

Carrots and Sticks

At the dawn of President Jimmy Carter's new administration, *Small Is Beautiful* was making a splash in the United States. After Schumacher's first American speaking tour coincided with the 1973 oil shocks, his book began to sell "like hotcakes."[1] He returned to the United States in March 1977, where he spoke on a conference panel at the University of Illinois, Chicago. A few days later, on March 22, he was invited to the White House, where at 1:45 p.m. he met with President Carter.[2] They spoke for nineteen minutes and posed for press photos. By the end, he'd won the new president over to the "small is beautiful" mindset. The conversation spurred many policy changes within the US government to promote appropriate technology programs. In the fields of medicine and international development in particular, where critics of modernization and medicalization pointed to the unavailability of key technologies and the overreliance on others, the tantalizing idea that there was an "appropriate" level of technology that would lead to optimal social and economic development permeated the new administration's program. This was especially true at USAID, where appropriate technology was put forward as a commonsense solution both to reports that aid was not reaching the rural poor and to the agency's budget shortfalls.

American interpretations of which technologies were appropriate for the developing world dominated discussions of international development largely because the donor funds the US wielded bilaterally dwarfed what it made available through multilateral UN organizations by a ratio of about four to one.[3] At a series of conferences on international development throughout the 1970s, US policy on technology transfer to the developing world had outsized influence. Behind the scenes, shaping US policy was a long, arduous process, often hinging on trying to define appropriate technology in a way that could be realistically supported by the dwindling US foreign aid budget. The government shied away from specifying which specific technologies were "appropriate"—officials were loath to make any promises when budgets were tight and the private-sector companies that supplied many in-demand technologies were not necessarily cooperative. To that end, US policy makers could lean on the fact that appropriate technology was defined by its local suitability (and every locality had different needs) to make more general statements about technology selection in foreign aid. This came to a head at a UN conference in Vienna in 1979, although the lead-up to the event was perhaps more significant in defining US policy toward technology transfer to the developing world.

∵

On August 20, 1979, after years of preparation, Father Theodore Hesburgh stepped up to the podium in Vienna's Stadthalle, a massive convention center built in the mid-1950s by the modernist Austrian architect Roland Rainer, to deliver an opening statement about the US position to the UN Conference on Science and Technology for Development (UNCSTD).[4] "It is only proper that this beautiful city by the Danube, a witness of so many great historical events, should be the site to compose the new contours of our future," he began.[5] After some initial introductions, he went on to express his reservations about the newly independent African and Asian states' focus on science and technology. His words were reminiscent of Schumacher, by the end of his life a fellow Catholic, in their warning against materialism and their appeal to a higher meaning: "Science and technology are not the guarantors of civilization; they only guarantee the possibility of civilization. Fast cars or fast breeders, synthetics or cybernetics, do not a civilization make. Unless our existence reaches beyond the frivolities of materialism and becomes a life enriched with meaning, science and technology will not be hallmarks of progress; they will only be the trappings of modernity. The pursuit of scientific excellence must be based upon the pursuit of human goals."[6] Positioning science and technology as necessary but not sufficient for development, his words can be read as an attempt to temper developing country expectations. It was a carefully crafted approach.

USAID was in the midst of a budgetary crisis. In the 1960s, the agency had funded many large-scale, high-capital projects like dams, manufacturing plants, and highways, according to the logic of modernization theory. When the tide of popular opinion turned against foreign aid in 1968—the United States having sustained heavy casualties during the Tet Offensive in the Vietnam War—a current of critique arose against the modernization paradigm. That fiscal year, USAID's budget was cut by nearly 26 percent. Things got worse after the 1973 oil shocks, which—together with the mounting costs of the war—dealt the agency a further 15 percent cut. Many measures had been introduced to try to close the budget gap. The 1973 New Directions policy signed by President Nixon as Public Law No. 93-180, an amendment to the 1961 Foreign Assistance Act, is most often cited as the law which shifted USAID's focus from elite modernization projects to those focused on the poorest citizens—consonant with the "basic human needs approach" that would be taken at the World Bank a year later and in line with the spirit of the New International Economic Order (NIEO), a coalition of newly independent states advocating for structural change at the United Nations. However, the legislation also put forward that "US cooperation in development should be carried out to the maximum extent possible through the private sector."[7] This led to significant changes within the project management structure of USAID, which had previously held much of its technical expertise in-house. Contracting out to "private and voluntary organizations" expanded the capacity of the cash-strapped agency.[8] Appropriate technology went hand in hand with this—it was cheap compared to big modernization projects and could be contracted out to small firms and nonprofits.[9]

While USAID was struggling to make ends meet, the US delegation sought a way to temper expectations and commitments for costly new technologies. The Vienna conference delegates prepared for UNCSTD for over two years, strategizing how best to save face with the developing world. Their preparatory committees (often referred to as "PrepComs") reveal how these Carter-era policy makers navigated the shrinking budgetary and popular support for foreign aid domestically and the expanded demands for primary health care, technology, and essential drugs internationally. As at USAID, appropriate technology again emerged as a possible solution.

Father Hesburgh, a Catholic priest and then president of the University of Notre Dame, was appointed by President Carter to lead the US delegation. He was an experienced civil servant. President Eisenhower had appointed him to the National Science Board in 1954, and he had served on several presidential commissions from that point on, as well as on the board of the nonprofit Rockefeller Foundation. Although he had agreed to base a pilot for President Kennedy's Peace Corps at Notre Dame in 1961, his primary commitment throughout the 1960s and 1970s was to the civil rights movement at home.[10] He led the US Civil Rights Commission from 1957 until he was dismissed from his post by the Nixon administration in 1972, framing civil rights as a moral issue that people of faith should rally behind. In 1961, dissatisfied with the Kennedy administration's record on civil rights, he wrote in the commission's annual report: "Personally, I don't care if the United States gets the first man on the moon, if while this is happening on a crash basis, we dawdle along here on our corner of the earth, nursing our prejudices, flouting our magnificent Constitution, ignoring the central moral problem of our times, and appearing hypocrites to all the world."[11] This statement, made well before his appointment to the head of the US delegation to UNCSTD, is telling. It suggests that Father Hesburgh understood both the distraction that high technology could be to more pressing issues like human rights and the morally tenuous position the US occupied as the leader of or model for international development efforts.

And yet he took the post seriously. As Father Hesburgh looked forward to prepare for the conference, he was careful to consider past US efforts on science and technology for development. The United States had come belatedly to this game—until the 1960s, the United States did not have a well-established policy or office for facilitating the exchange of scientific information with developing nations.[12] This changed in 1961, when, in the wake of the highly successful 1955 Atoms for Peace conference, then UN secretary-general Dag Hammarskjöld and his Science Advisory Committee proposed a conference on how science and technology could be used to benefit countries in the less developed world.[13] This became the 1963 UN Conference on the Application of Science and Technology for the Benefit of the Less Developed Countries (UNCAST), the predecessor to UNCSTD, held under the new administration of Secretary-General U Thant.

One of Father Hesburgh's advisers, W. Murray Tood, executive director of the Commission on International Relations at the National Academy of Sciences, asked Dr. Walsh McDermott for his reflections on the earlier conference. McDermott had chaired the American effort at UNCAST and was also on the advisory committee of the US Vienna delegation. His candid response contrasted sharply with the official US postconference report out of the State Department and, as a result, was in many ways more useful to the delegation.[14] McDermott surmised that a major impetus for the conference came from the Soviet delegates on Hammarskjöld's committee, who, following on the success of the Sputnik in 1957 and Gagarin's orbit in space in 1961, saw an opportunity to showcase their perceived superiority in science and technology to what was then called the Third World.[15] In McDermott's retelling, the US government institutions responsible for science policy and international organizations were caught off guard by the conference effort and he was asked to step in to develop the US position because of his former work with the president's Science Advisory Committee. He took the job to prevent "a Soviet triumph."[16]

Ahead of UNCAST, the Americans quickly set up structures to determine US policy on science and technology for development. McDermott established a "Public Advisory Board" within the Department of State, which consisted of a number of well-known American scientists from industry and academia, including the geophysicist Frank Press of CalTech, the nuclear physicist Jerrold Zacharias of Massachusetts Institute of Technology, and J. Herbert Hollomon of General Electric. Representatives of the Rockefeller Foundation, including President George Harrar and director of Medical and Natural Sciences Robert Morison, and development economists, including Max Millikan of MIT and Frederick Harbison of Princeton, were also part of the board.[17] The foreign secretary position of the National Academy of Sciences, which had, according to McDermott, effectively been in "abeyance" when the conference was announced, was swiftly filled by Harrison S. Brown, a geochemist at CalTech who had worked with Zacharias on the Manhattan Project. By February 1963 when the conference kicked off, McDermott felt that the Americans were on a more level playing field with the Soviets—John Glenn had orbited Earth in 1962, demonstrating the capability of US space technology, and the US delegation to UNCAST consisted of one hundred "high caliber" people, many of whom were Nobel laureates.[18]

The UNCAST conference itself acted as a showcase for US and Soviet science and technology but had little appreciable impact for the developing countries it was intended to benefit. Partway through the conference, delegates from developing countries organized what McDermott described as a "gripe session," airing their discontent with the conference to that point. The two chief grievances were that they resented being witness to the "US and Soviet verbal counter-punching" (a concern McDermott thought was overblown) and that, with the large conference sessions, they did not have the opportunity to learn much from the presenters. In response, the US delegation organized a "shadow conference" for which they booked smaller rooms at the conference facility and hotel rooms around Geneva to host a series of small, informal meetings with the delegation scientists. The shadow conference, in effect, was one of the few informal outlets for the exchange of ideas. As McDermott describes:

> It did do the one thing that people always talk about with respect to conferences, namely that it isn't what happens in the meeting rooms, it's what happens in the corridors. In this case, with the wide range of subjects and the wide differences both culturally as well as nationally, what went on in the corridors wasn't worth a damn. Indeed, it was not until the actual shadow conference was set up that there really was much interchange along these lines. To be sure, there was the inevitable reception every evening at which there was considerable interchange back and forth. How much of this was truly professional, however, and how much

> consisted of the usual stilted conversations one has in Esperanto on such occasions, is hard to say.[19]

His approach proved successful and won favor for the US delegation at the meeting itself; however, it did not translate to any form of action after the meeting concluded—a fact that was highly criticized in 1976 when the call for science and technology for development was renewed at the United Nations by the NIEO. Father Hesburgh therefore faced considerable pressure from the Carter administration to find solutions that would make America look good in the eyes of the developing world.

The formal resolution that the Vienna conference would be held was passed in December 1976. It was one of several United Nations–based NIEO initiatives aimed at improving the status of developing nations.[20] US preparations for UNCSTD began relatively early, in 1977, when ambassador to the UN Jean Wilkowski established an office for the conference within the State Department. Wilkowski was one of the first women to become a career diplomat in the US Foreign Service and was appointed to the UN ambassadorship fresh from her posting to Zambia, where from 1972 to 1976 she was the first woman to serve as an American ambassador to an African country. Originally planning to become a journalist, she had joined the Foreign Service in 1944 after being persuaded that the US government needed stronger "morality and ethics" in its conduct of international relations by an American Catholic priest and historian who had traveled through South America.[21] Her sense of morality, and her Catholic roots, meant that she became a natural ally of Father Hesburgh's, and the two worked closely together to develop the US position for Vienna.[22]

As had been the case before UNCAST, US policy on many issues related to UNCSTD was ill defined. The preparatory committees as well as congressional hearings and a number of smaller conferences and workshops served to build up the US position. The preparatory meetings were attended by those in the American delegation to the Vienna conference as well as outside advisers and interest groups. Among the US delegates was Dr. Guyford Stever, a physicist with a PhD from CalTech who had previously served as associate dean of engineering at MIT, president of Carnegie Mellon, director of the National Science Foundation, and as the first director of the Office of Science and Technology Policy under President Ford. His papers from the meetings, as well as from the conference itself, illuminate both the breadth of the American effort to define its policy on science and technology ahead of UNCSTD as well as the uncertainty over how to react to what they expected the NIEO and Soviet bloc's demands to be.

Appropriate technology featured heavily in the US strategy. Father Hesburgh and Ambassador Wilkowski focused on leading with carrots rather than sticks. During the advisory meetings, the delegates discussed many different ways to entice developing countries over to the American interpretation of appropriate technology—that which was relatively low-tech, inexpensive, and nonthreatening to patent protections. For countries in the Middle East, for example, UNCSTD was discussed in the context of the newly signed Camp David accords.[23] The peace process could be enhanced, the delegates thought, through US science and technology, especially in the areas of natural resources, agriculture, industry, health and nutrition, pharmaceutical manufacturing, and infrastructure—provided that Israel and Egypt continued to cooperate in good faith.[24] Other contextual factors ("leverage") that were "amenable" to Western (especially American) influence were ranked in a background paper for the delegates, "The Role of Technology in North/South Relations" by Michael Oppenheimer and John Pothier of the Futures Group.[25] The most important of these, the authors noted, were direct aid initiatives and bilateral talks, and indirect influence over "delegate backgrounds, political/symbolic needs, secretariat activities, and the preparatory process."[26] They also identified several "wild cards" for the conference, which included the "intrusion of peripheral political disputes" (Cambodia and Vietnam within the Group of 77, the Soviet Union and China, Argentina and Chile), the change in orientation of "key" developing countries (Algeria, Brazil, Egypt, Kenya, Nigeria, Venezuela, and Yugoslavia), potential worldwide crises ("persuasive evidence of climate change," major natural disaster, major nuclear accident), and the "posturing" of challengers to replace the outgoing UN secretary-general Kurt Waldheim, whose term was to expire at the end of 1981.[27]

Within the official US delegation briefing book, the locus of bilateral assistance offered at the conference was to be USAID, a newly proposed Institute for Scientific and Technical Cooperation, Appropriate Technology Incorporated, and the Overseas Private Investment Corporation.[28] Appropriate Technology Incorporated, a private and nonprofit corporation, was established in 1976 by Congress to "promote the development and dissemination of technologies appropriate for developing countries."[29] It provided small grants to private-sector businesses in the developing world to help them develop and promote appropriate technologies on their local markets. The Overseas Private Investment Corporation served US businesses interests—it was founded in 1971 to enable privately held American companies to invest, with limited risk, in developing country markets.

Father Hesburgh was pragmatic. He believed that the world in 1979 was "interdependent"—"we need [less developed country] markets and natural resources and [less developed countries] need our technology. There should be some way in which we could get together to work out some of the problems to our mutual interest."[30] The solution for the US delegation, he proposed, was to focus on the areas of basic human needs, including hunger, malnutrition, poverty, and disease, which could be ameliorated through appropriate technology.[31]

∴

As the preparatory committee meetings were underway, appropriate technology policies also came under US congressional oversight. Major cuts to the US foreign aid budget, rather than adherence to any sort of neoliberal economic theory, had been the main driver for adopting appropriate technology programs. Influenced by Schumacher and VITA, appropriate technology was perceived as more cost-effective than large-scale modernization efforts within USAID in the early and mid-1970s, and a pragmatic way around shrinking program budgets. Following this initial push, in July 1978 the Subcommittee on Domestic and International Scientific Planning, Analysis, and Coordination of the Committee on Science and Technology of the US House of Representatives convened hearings on appropriate technology and its role in US bilateral foreign assistance.[32] Based on the opening statement of Congressman James H. Scheuer (D-NY) who chaired the subcommittee, the impetus for the conference seemed to be the increased skepticism of many policymakers and development experts toward the large-scale capital-intensive projects built in the name of modernization theory. He attributed this to a "complex of concerns" arising from, among other events, the environmental movements of the 1960s and early 1970s, the oil shocks in 1973, and "disenchantment with many of the massive welfare programs of the 1960s (as embodied in the Great Society)."[33] The hearing called witnesses from academia, such as development economist Dr. Gustav Ranis of Yale University—who was also part of the UNCSTD Preparatory Committee, nonprofit foundations, and other government departments, including USAID.

The purpose of the hearings was to explore the "appropriate technology movement" in foreign aid and how it could be used to meet the "basic human needs" of the world's poor, a phrase borrowed from the very similar World Bank effort under its director Robert McNamara. Congressman Scheuer, and his "star witness" and former teacher, economist and Congressman Clarence D. Long (D-MD), advocated strongly throughout the questioning for all manner of "appropriate technologies" in health, including village midwives and community health clinics, consistent with a broad-based, primary health care approach.

Yet this enthusiasm did not extend to technological areas controlled by US business interests. Contiguous with the preparations for Vienna, the US government was in the midst of negotiations within the UN Commission on Transnational Corporations, a subcommittee of the Economic and Social Council, to establish a code of conduct for transnational corporations in the developing world—another NIEO effort.[34] While the Soviet bloc and Group of 77 viewed the purpose of the code of conduct as a means to exert more control over transnationals through a legally binding agreement, the United States and other Western nations pushed for a voluntary code of conduct that stressed standards of acceptable behavior both for the corporations and for the developing country governments hosting them. A background paper specifically on the issue of transnational corporations was prepared for the US delegates to Vienna so that they could speak to the issue in a coordinated way. The administration feared that technology transfer agreed upon through UNCSTD, without a robust agreement protecting American transnational corporations' interests in G-77 countries, including nondiscriminatory treatment of foreign enterprises, respect for contracts, and standards of expropriation and compensation, would undermine American businesses in an already-precarious economy.[35]

One way around this was to precisely define which forms of technology were appropriate for transfer. At the hearings, Dr. Ranis presented a prepared statement, "Appropriate Technology: Obstacles and Opportunities," which defined appropriate technology as the "optimum choice" of technology that would lead to the "maximization of societal objectives given the society's capabilities."[36] Total optimization required, in theory at least, perfect knowledge of the market, which Ranis conceded was perhaps not appropriate in the case of information protected by patent rights—as in the pharmaceutical industry. Ranis suggested that, to restore perfect competition as much as possible, international patent legislation be revisited to "ensure that patents are actually used (and technology transferred) rather than serve as a way of controlling markets and inhibiting the flow of information."[37] In the discussion that followed his address, he clarified that although "core technologies" were unlikely to be adapted to developing countries' needs because of patent restrictions, appropriate technology could be adapted for "related operations" such as packaging pharmaceuticals (something already done, to some extent, by developing country subsidiaries of multinational corporations).[38] This answer seems to have satisfied the subcommittee, as patent rights were not raised again throughout the three-day hearings.

The subsequent silence on patents, when the goal of meeting basic human needs would arguably be undercut by the exclusion of "core technologies" such as pharmaceuticals in the appropriate technology mandate, could also be read as part of a larger deference to, and increasing dependence on, private industry in bilateral foreign aid. With rising balance of payments problems, the United States was also keen to keep money within the US economy and had no interest in risking the intellectual property (and profits) of US companies. The New Directions policy had propelled the involvement of the private sector in foreign aid, but deference to industry on technology transfer solidified their influence on policy.

Sander Levin, then assistant administrator of USAID, testified during the last day of the hearings. In the prepared statement, he relayed the agency's commitment to appropriate technology programs and reported that "the rate of introduction of the appropriate technology concept into AID projects has been accelerating so that it now permeates all programmatic planning."[39] In the ensuing discussion, he stated his belief that the agency's New Directions policy "has helped to begin to move the pendulum" toward a focus on the basic human needs of the poorest citizens rather than one on high-capital projects that primarily benefited developing country elites.[40] Nevertheless, Congressman Scheuer asked why there had not been more of an impact, which sparked a debate over the boundary between appropriate and what he called "high technology." Mr. Levin responded that while USAID was beginning to see an impact, the agency wanted to be sensitive not to "dictate" the needs of developing countries. Congressman Scheuer shot back:

> It seems to me—perhaps not in your agency, but there has been too much of this pusillanimous and faint-hearted feeling in the State Department that we cannot afford to ruffle any of our mission Embassies, and our Ambassadors cannot afford to ruffle anybody's feathers by raising these questions . . . I am sure you are totally aware that there is not a lot of public support for foreign aid . . . It takes a lot of guts for a Congressman to vote for foreign aid today. And if they feel that these programs are not fine-tuned to help people . . . you will find that support approaching a vanishing point.[41]

He went on to question USAID's preparation for UNCSTD, which by that point had been underway for two years, and insisted on confirming that US policy would be coherent across that conference as well as two others on industrial technology.

As was made clear throughout the congressional hearing, American interpretations of appropriate technology focused on solutions that were simple, small-scale, low-cost, and off-patent. In short, these were exactly the opposite of the "high" technologies that many developing countries wanted. USAID attempted to steer the conversation toward the much more circumscribed American vision of appropriate technology more directly. The agency set up a little-known program to "assist" developing nations with the preparation of their UNCSTD position papers. It aimed, explicitly, to "increase US influence in the Third World, provide us with direct information on LDC [less developed country] preparations for UNCSTD, and contribute to a responsible third world position at the conference."[42]

It is difficult to assess the impact of this program—there is evidence that only one country, Mauritania, opted to receive the assistance.[43] And why?[44] For one thing, Mauritania wanted to maintain its close relationship with France, its former imperial power.[45] France, also under pressure from industry, sided with the United States on the question of appropriate technology. For another, Mauritania had frequently leveraged its relationship with France to bring in "expatriate specialists" to bolster its domestic industries, as in 1963, when it needed to revitalize a declining mining industry.[46] The real crux of Mauritania's motivation may, however, be that in the wake of the 1973–1974 famine, Mauritanian government resources were stretched so thin that foreign voluntary agencies of all stripes increasingly stepped in to fill the gaps and govern the nation, and by 1978, preparing for a large international conference was just one other area in which the government felt comfortable handing the reins over to a foreign volunteer.[47]

Assistance to Mauritania was contracted by USAID to the International Science and Technology Institute, a US-based private consulting group that had sprung up the year before in the midst of the Vienna conference preparation frenzy. With the major cuts to its budget, USAID relied more and more on private contractors and nongovernmental organizations to carry out its projects in this period—an interesting parallel to what historian Gregory Mann called the increasing "nongovernmentality" concurrently happening in West African governance.[48] The project was allocated $35,734 for two expert consultants to go to Mauritania for a month to assist the government in preparing for UNCSTD. According to the contract, USAID required the consultants in Mauritania to "direct advisors' attention to AID's mandated interest in assisting the poor majority, providing for basic human needs," and to report back to Washington on "the nature of [the] host country's UNCSTD preparations and on host country's principal concerns in relation to conference topics."[49]

The consultants arrived in the capital, Nouakchott, in June 1978, just one month before a military coup. While they had originally planned to work with counterparts in the Mauritanian government—domestic experts in the areas of science and technology, agriculture, and industry—"this proved impossible due to other commitments of government staff."[50] Instead, the team was "advised by the GIRM [Government of the Islamic Republic of Mauritania] 'to be creative' and not to be limited or bound by the UN Guidelines for preparing papers for UNCSTD."[51] The resulting papers took as their goal "the selection of appropriate science and technology" for solving critical Mauritanian problems pertaining to drought, desertification, and food shortages.[52]

The consultants perceived the situation to be so urgent and acute that they wrote that "to wait for the time-absorbing process of training scientists and the initiation and conduct of research to find solutions, and then to put them into effect could be disastrous. The damages inflicted by such delays may be almost irreversible. In consequence, short cuts must be taken through identifying appropriate technologies for conserving and managing resources for improving agricultural production."[53] Appropriate technologies that the consultants suggested included mulching, improved crop varieties, pesticides, irrigation practices, and veterinary medicines—interventions already widely employed in agricultural practice in the region. Furthermore, they warned that "other technologies and expensive capital facilities need more cautious evaluations."[54] They stressed, in an appendix to the papers, a reliance on "non-proprietary technology," what they describe as "the simplest and most inexpensive system" and that included expired patents.[55]

The reason these nonproprietary technologies are not more used by developing nations, the consultants claimed, is that there is a gap between knowledge of the technology and the end user. The solution is therefore not to invest in more high-capital technology or research and development, but rather in an "Appropriate Technology Research Institute" that would act as a "Technology Delivery System" for end users. Emphasizing the community development paradigm employed in Burma and India, "model villages" would be established as a "showcase" for the technology transfer of appropriate interventions, thereby providing other villages with an example to emulate and spreading knowledge of these simple technologies "just by 'osmosis.'"[56] The consultants directed the Mauritanian government's priorities—quickly and urgently—toward low-cost, low-tech interventions and away from technologies protected by US patents.

The USAID assistance program to developing countries for the preparation of their UNCSTD position papers is not surprising. As the United States has historically delivered between 70 percent and 90 percent of its foreign aid bilaterally (mostly channeled through USAID), as opposed to through multilateral organizations such as the United Nations or World Bank, it is fitting that the US delegation would seek to use its bilateral ties to influence a large multilateral forum of significant importance, such as the Vienna conference. In the case of Mauritania, it is likely that with an overstretched government still recovering from the 1973–1974 famine and with an interest in maintaining amicable relations with France and its Western allies, the US assistance program was appealing even if the position contrasted with the more forceful one staked out by the NIEO. The military coup that occurred just as the consultants left Nouakchott does not appear to have affected the USAID program—it was not mentioned in the final report, which was published after the change in regime.[57] Mauritania remained a "responsible presence" for the United States at the Vienna conference, and its presence allowed the US delegation to claim that it was not alone in advocating for a more restrained definition of appropriate technology.

∵

Pharmaceuticals were often invoked as an example during policy discussions of appropriate technology in the lead-up to the Vienna conference, an issue of particular concern to the American pharmaceutical industry. As in Dr. Ranis's testimony and in the proposed assistance to Israel and Egypt after Camp David, access to pharmaceutical technology could be both a carrot and a sticking point. In January 1979, Senators Kennedy, Javits, and Schweiker of the Senate Subcommittee on Health and Scientific Research—who believed that neither the US pharmaceutical industry nor biomedical research laboratories in US academic institutions were devoting sufficient time to address the health problems of the developing world—held the conference "Pharmaceuticals for Developing Countries." It was sponsored jointly by the Institute of Medicine; the Department of Health, Education, and Welfare; and the Pharmaceutical Manufacturing Association. The Institute of Medicine, now the National Academy of Medicine, was a nongovernmental organization that nevertheless held a congressional charter, whereas the Pharmaceutical Manufacturing Association is an industry lobbying group. Participants included members of industry, government, academia, and the nonprofit sectors from both developed and developing nations. They discussed the development, availability, and accessibility of drugs for the major diseases that burdened developing health systems and the challenges the pharmaceutical industry had in playing a larger role in overcoming them. A key debate emerged over the use of pharmaceuticals as appropriate technologies in broad, horizontal health systems (as in the WHO's essential drugs program and primary health care) versus the use of pharmaceuticals as specialized weapons in narrow, vertical disease campaigns, which enabled the American industry to maintain more control over their deployment.[58]

In his keynote address, Senator Edward Kennedy affirmed his support for access to essential drugs and the provision of primary health care.[59] While this set the tone for the conference as a whole, many of the presentations that followed were focused on the problems and constraints of drug delivery, or ways that programs could be reformed to be more cost-efficient. A panel presentation by Dr. Kenneth Warren of the Rockefeller Foundation, who would go on to coauthor a highly influential paper on selective primary health care three months later, is illustrative. While he argued for greater cooperation between academe and the pharmaceutical industry, his primary example of how this cooperation could function was how greater study of schistosomiasis enabled chemotherapeutic campaigns to reasonably treat fewer people and with lower doses of the drug and thus be more cost-effective.[60] The example, while nominally in support of primary health care, reduced the complex problem of schistosomiasis to a single, narrow technological intervention (chemotherapy) just as selective primary health care significantly narrowed the scope of primary health care services outlined at Alma-Ata.[61]

Dr. Walsh McDermott, the former chairman of UNCAST who was then working for the Robert Wood Johnson Foundation and as a special adviser to the president, was another prominent speaker. In his address on the historical perspective of pharmaceutical development and distribution, he engaged with the debates over interventionism that were raging in parallel in foreign aid and in medicine itself. He challenged the controversial study of the British physician Thomas McKeown that argued that death rates from tuberculosis and other common diseases of the nineteenth and twentieth centuries fell primarily as a result of improvements in social, economic, and environmental conditions, rather than scientific or technological advances in medicine.[62] McKeown's thesis was taken by advocates of the basic human needs approach as proof of the efficacy of the horizontal primary health care concept. However, McDermott argued that, by putting McKeown's data to a logarithmic scale, the effect of pharmaceutical intervention on death rates (in this specific example, from tuberculosis) was both significant and clear—or, as he put it, "the technology works."[63]

While McDermott was undoubtedly an interventionist and held a strong belief in the promise of medical technology, his presentation was not an unqualified approbation of vertical health programs. He was attentive to context, and he recognized that successful pharmaceutical intervention for many developing country diseases required the improvement of sanitary conditions. Nevertheless, McDermott advocated for a new system of pharmaceutical development "especially tailored to both the financial and biological needs of the problem," and closed by expressing hope (with the adage, "sometimes the poor get lucky") that pharmaceutical intervention would overcome the obstacles of poverty.

In a brief statement intended to describe the administration's perspective on international health, Dr. Gilbert S. Omenn, an associate director at the Office of Science and Technology Policy of the White House who represented the Carter administration at the Institute of Medicine conference, attested to President Carter's deep commitment to meeting the basic needs of developing country populations through community-based primary health care. He indicated that USAID formed the "core of United States government efforts to promote better health in the Third World" and had been allocated $130 million in fiscal year 1979 for health delivery programs, environmental health, disease control, and health planning.[64] He also outlined President Carter's plan, first announced in a speech in March 1978 in Caracas, for the creation of an Institute for Scientific and Technical Cooperation. This was to be a research and development institute focused on the application of science and appropriate technology to developing country issues, particularly in the areas of health, agriculture, and energy production. The Carter administration's international health policy also included more cooperation with US universities and private and voluntary organizations, as well as an increased role for the private commercial sector, augmenting the New Directions policy in place since 1973. Omenn concluded by appealing to the participants at the Institute of Medicine conference, from the pharmaceutical industry, academia, and foundations, for "greater recognition of human needs."

The Institute of Medicine conference served as a venue, just after Alma-Ata, in which US policy toward primary health care and essential drugs was tested by industry and consultants. Although Senator Kennedy and Gilbert Omenn, on behalf of President Carter, showed strong support for both concepts, the majority of presentations stressed reservations about the feasibility or desirability of such an approach. Many proposals offered technical, vertical solutions, and appropriate technology was considered, much as it had been by Gustav Ranis at the congressional hearings on appropriate technology, for the realm of packaging and delivery systems rather than for core pharmaceutical technologies. Moreover, the concept of essential drugs—key to primary health care as defined at Alma-Ata—was invoked only three times: once by a representative of the World Health Organization (WHO), who was supportive; once by William Hubbard of the Upjohn pharmaceutical company, who associated it with communist-bloc health assistance; and once by the director of the German pharmaceutical company Farbwerke Hoechst, who expressed deep skepticism of the WHO's plan to provide basic drugs at significantly reduced cost to countries most in need.[65] It is not surprising that, given the political climate in the US toward foreign aid and the industry reactions to primary health care at the conference, a few months later at another conference on implementing primary health care in Bellagio, Italy, selective primary health care emerged as the leading international health strategy.

This focus on pharmaceuticals meant that USAID soon had to articulate its own policy toward their use in international development ahead of the Vienna conference. The agency commissioned a report on the pharmaceutical industry's position on technology transfer to the developing world as an addendum to an earlier report for the conference on "patterns of accommodation by US firms to emerging [less developed country] demands for new modes of technology acquisition."[66] The contractor that prepared the report, Developing World Industry and Technology, was the private consultancy of Jack Baranson, an economist with the World Bank who wrote several monographs in favor of technology transfer in international development.[67] The report, which positioned the US pharmaceutical industry as "one of the oldest protagonists in the ongoing debate between multinational corporations and newly industrializing countries over appropriate and advantageous technology transfer," expressed pharmaceutical industry concerns over technology transfer, including patent rights, the time invested in research and development, and adequate compensation.[68]

USAID had also sought the advice of a panel of experts assembled by the Institute of Medicine, which delivered its committee report "Review of the Agency for International Development Health Strategy" in September 1978. Reflecting more broadly on USAID program guidelines in the health sector that were developed in the lead-up to UNCSTD, the committee questioned USAID's focus on primary health care, in part given the lack of availability of drugs and vaccines in many countries. The committee comprised experts in tropical medicine and international public health from academia, multilateral organizations, and nonprofit foundations, including Dr. Donald Henderson, then dean of the Johns Hopkins School of Public Health. Perhaps unsurprisingly, given Dr. Henderson's renown for his role in leading the WHO's successful smallpox eradication campaign, the report recommended that USAID proceed slowly in implementing primary health care projects—what Henderson referred to as the "doc in a box" approach (the "box" being the rural clinic)—and not turn its back on the narrower targeted disease campaigns.[69] Quoting a 1966 article of Dr. Walsh McDermott's, the report argued, "The biomedical goal of international development, or purposeful modernization, is to modify the disease pattern of an overly traditional society to a disease pattern that will not act as a major drag on a modernization effort."[70] Targeted disease campaigns, which at that point had proven track records, therefore must take precedence over unproven primary health care strategies.[71] This was especially true, the committee went on, in developing countries which lacked "effective management . . . programs for training individuals for performing the requisite tasks . . . communication systems and facilities for patient referral . . . [and] basic distribution systems to assure that drugs, vaccines, and other required commodities reach the point of delivery regularly and in satisfactory condition."[72] It was a bit of a catch-22. Without infrastructure to support primary health care, it could not be successful. But channeling funding toward disease-specific vertical campaigns designed to overcome this meant that money went to targeted, often pharmaceutical, interventions and investment in the needed infrastructure never came.

∵

Ahead of UNCSTD, therefore, the USAID health strategy was caught between two competing factions. On the one hand, Congress favored low-cost appropriate technology and primary health care consonant with multilateral policy. On the other, the pharmaceutical industry preferred disease-specific campaigns and tighter controls on drug distribution for a variety of economic and logistic reasons. In the end, the Preparatory Committee health panel decided to reframe appropriate technology in a way that made it more palatable, and less threatening, to industry interests. Rather than commit to any particular technologies as "appropriate" for transfer, the US position emphasized its support for research and development specifically for the developing world through the ISTC. As policy makers worked out the US stance on technology transfer ahead of the Vienna conference, multilateral bodies were busy with their own set of preparatory meetings. There, newly independent states asserted a very different view of what "appropriate technology" meant.

[CHAPTER FIVE]

Visions of the Future

As the Americans hashed out their approach to bilateral technology transfer ahead of the Vienna conference, their talks hinged on appropriate technology—the idea that some technologies were suitable for transfer to developing countries while others were not. It was also a major topic of discussion within the major multilateral organizations working in international health.[1] Here, too, a series of conferences helped to hone the organizations' approach.

The World Health Organization's (WHO) 1978 Alma-Ata conference on primary health care is most frequently invoked by historians of international health. Appropriate technology and essential drugs were key pillars of the WHO's vision of primary health care. Alma-Ata was followed, a few months later in April 1979, by a smaller conference in Bellagio, Italy, "Health and Population in Developing Countries," organized by the Rockefeller Foundation, which discussed the feasibility of implementing primary health care as detailed at Alma-Ata. That conference ultimately resulted in the drafting of "selective primary health care," known for its selective focus on the four interventions of GOBI (growth monitoring, oral rehydration, breastfeeding, immunization).[2] There were also a series of smaller, regional conferences in the lead-up to the 1979 Vienna conference that deserve attention—notably the Conference of Ministers of African States Responsible for the Application of Science and Technology to Development, or CASTAFRICA, in 1974—as they helped to define policy for the New International Economic Order and, thus, the intended recipient nations of international health programs.

While the WHO focused on primary health care, the World Bank similarly focused on programs that would meet the "basic human needs" of the world's poor. The NIEO advocated for science and technology transfer to enable its members to develop more rapidly. Appropriate technology seemed to be a concept around which these different constituencies and the American bilateral aid program administered through USAID could all rally. However, this appearance of alignment masked fundamental disagreements over what "appropriate" technology actually meant in practical terms.

The World Health Organization tends to get the most attention when it comes to the organization's stance on appropriate technology in international health. Many accounts positioned appropriate technology as an ancillary movement that was part of the larger turn toward primary health care at the Alma-Ata conference.[3] Others have appeared to credit the WHO's director general Halfdan Mahler with the creation of the appropriate technology concept and viewed the adoption of an appropriate technology mandate as a form of strategic adaptation for the organization caught between the Global North and Global South in the late 1970s.[4] Both versions positioned Alma-Ata as a transformational, idealistic moment that failed, and the NIEO at the United Nations as a movement in which valiant postcolonial southern nations gained a voice on the international stage but then were undermined by the exigencies of neoliberal economics.

Looking at the introduction of appropriate technology programs earlier in the 1970s, it is clear that the concept was adopted primarily for pragmatic and cost-saving reasons, rather than ideological ones. However, appropriate technology brought with it a focus on cost-effectiveness and the increased involvement of the private sector, which scholars would retrospectively identify as neoliberal tendencies. At the same time, appropriate technology can be thought of as the last gasp of community-oriented development, as the focus shifted from an egalitarian vision of global solidarity to one of meeting the basic human needs of individuals.

That these narratives played out most prominently in Geneva is perhaps not accidental. The city has an important place in the history of global neoliberalism. The historian Quinn Slobodian has shown how what he calls "Geneva School" neoliberals were key institution builders.[5] They used a series of Geneva-based organizations, from the League of Nations to the General Agreement on Tariffs and Trade, to insulate world economic markets from the democratizing demands of newly independent and decolonizing nations. This culminated in the formation of the World Trade Organization (WTO) in Geneva in 1995. These organizations were a scale up from the nation state, which aligned with neoliberal thinking that there was only one world market that unrestrained national democracy threatened. Decisions made at the multilateral level and enforced by rules and laws helped to preserve markets from unwelcome, rights-based calls for global equity. In limiting the pool of technologies available to the developing world and encouraging the privatization of aid, appropriate technology was leveraged as a practical way of maintaining the status quo and insulating markets (and market protections, like patent rights) from the demands of the NIEO.

As appropriate technology went from a development solution to *the* development solution for an influential group of people and publications, some began to wonder: what does appropriate mean? In meetings in Dakar, Washington, and Vienna in the mid- to late-1970s, the term *appropriate technology* was defined similarly but interpreted very differently. Although what happened at the WHO in Geneva is important, it is just one in a series of nodes crucial to the definition of appropriate technology. These places were not peripheral spokes to Geneva's center, as current historical accounts may suggest. The political and economic context of each gathering and of the places each delegate represented were critical to the reimagining of appropriate technology.

While US policy makers wielded outsized influence in defining appropriate technology in transnational aid, appropriate technology was also a locus for resistance. Delegates to international conferences from newly independent states in Africa and Asia rejected Western versions of appropriate technology that positioned them as poor, helpless, and bewildered by complex technology.[6] Postcolonial leaders, such as Léopold Sédar Senghor of Senegal, articulated modern, technologically advanced visions of the future. The reinterpretation of what was appropriate for the Global South was crucially important not only in the context of multilateral programs in the 1970s, but also, as we will see, in setting the groundwork for indigenous appropriate technology programs and—within the ongoing antiapartheid and liberation struggles in Southern Africa—in building "self-sufficiency" and independence from colonial regimes.

∴

The NIEO was a radical reimagination of the global political and economic system. The political theorist Adom Getachew described it as a welfare-oriented form of "anticolonial worldmaking," in envisioning "an egalitarian global economy."[7] Its proponents saw the ways that postcolonial state building was compromised by the poorer countries' economic dependence on richer nations. Though fashioned from Marxist critiques of economic dependency, the NIEO nonetheless sought solutions within the frame of globalizing capitalist economics. It aimed to improve the terms of trade for its members through the equitable distribution and redistribution of "technological progress," which would redress "the remaining vestiges of colonial domination, foreign occupation, racial discrimination, apartheid and neo-colonialism in all its forms."[8]

The 1974 NIEO declaration at the United Nations built off of an important but ill-remembered conference more than three months before, the Conference of Ministers of African States Responsible for the Application of Science and Technology to Development. Held in Dakar January 21–30, 1974, the conference was a site at which the newly independent nations of Africa articulated their own interpretations of what *appropriate* meant in the context of their priorities for state building. This built a concrete framework the NIEO could later use for advancing technological equity.

From the steps of the Palais de L'Assemblée Nationale, a large, white midcentury modern building that was built in 1956 to house the general council of the Afrique-Occidentale Française and that became the seat of the independent national assembly in 1960, the CASTAFRICA delegates looked over the main square of Dakar. Known then as the "rond-point de l'Étoile"—the roundabout of the star, the meeting point of many of the city's main avenues, which radiate outward from the center—the space was home to the statue of Demba and Dupont, a World War I memorial erected in 1923 that featured a French infantry soldier with his hand resting paternally on a Senegalese *tirailleur*. Considered by many to be a monument to colonialism, the statue was moved in 1983 to the Bel Air Cemetery, and the "rond-point de l'Étoile" has since been renamed Place Soweto, in honor of the Soweto uprising in South Africa and the end of apartheid. Nevertheless, as the delegates filed into the Palais Nationale in January 1974, these reminders of the proximate colonial past—and their place at the center of the former French colonial administration—were ever-present.

Mr. René Maheu, then director general of the UN Educational, Scientific, and Cultural Organization (UNESCO), the primary sponsor of CASTAFRICA, opened the proceedings. He was a controversial figure. Maheu, a high school philosophy teacher in Normandy before becoming a cultural attaché at the French Embassy in London and an information officer in North Africa during World War II, was part of the midcentury French intellectual scene.[9] He had attended the École Normale Supérieure in Paris with classmate Jean-Paul Sartre, a close friend whose politics leaned further to the left than his own. Maheu joined UNESCO when it was established in 1946, and, after a "bitter" fight for succession with Indian development economist Malcolm Adiseshiah, became the organization's director general in 1962. He was committed to the role, and particularly the issue of raising adult literacy rates, which he viewed as essential to cross-cultural human understanding.[10]

Maheu's tenure at UNESCO spanned the period of peak decolonization in Africa and Asia.[11] Though committed to the universality of human rights in a theoretical sense, he did not explicitly align himself with the postcolonial movement. He was, fundamentally, conservative. The historian of education policy Phillip W. Jones wrote in 1988 that, "both within Unesco and outside it, Maheu was seen as a strong-willed, stubborn, even dictatorial figure; few liked him personally, but never is there expressed doubt that his leadership was of unique substance, vigour and determination . . . Maheu identified greatly with the organization. Exhibiting a de Gaulle-like egoism, his attitudes were of a 'l'état c'est moi' quality."[12] It was a quality that landed him in trouble at CASTAFRICA, where postcolonial leaders were not so receptive to his commanding demeanor.

In his opening address at CASTAFRICA, Maheu made several pronouncements about what he thought the delegates should focus on during the conference. On the issue of technology transfer, he said: "Not only does it entail decisions of a delicate political and economic nature regarding the source of imported technologies, but, in addition, it raises the basic question of suitability of these technologies. Every foreign technology which is not suited to the specific possibilities and needs of its users is destined to remain an alien element whose potential cannot be exploited as it should and whose entry onto the scene may even result in dangerous tensions and distortions in the general development of the country."[13] His invocation of technology as being potentially ill suited and dangerous for Africa did not endear him to an audience who came expressly to discuss its promise for their continent. Nor did the end of his address, in which he positioned lack of science, rather than decades of colonialism, as the root of Africa's underdevelopment: "For whereas culture is, admittedly, the source of your identity, only science, whose flowering it is the principal task of modern education to bring about, can, as you well know, at long last restore you to the intellectual and practical means of active participation in shaping the destiny of the world."[14]

His conduct at the conference did not improve over the course of the next few days. According to two American observers, midway through the conference, in an "intemperate lecture," Maheu reportedly "scolded" the delegates for criticizing UN organizations and their development efforts, which he said were the "best answer" for their needs. One delegate noted that "the reactions to Mr. Maheu's views were so strong that a special closed meeting was hastily arranged to soothe feelings, in spite of one delegation's call for a public apology for public insult."[15] Delegates felt that Mr. Maheu's attitude, and those of his French staff, reflected colonial sensibilities and not the new realities of the self-sufficient African future they envisioned.

Mr. Léopold Sédar Senghor, the first president of independent Senegal, gave his own address to the CASTAFRICA delegates, after Maheu's. Senghor was also educated in Paris, at the Lycée Louis-le-Grand and the Sorbonne, and, like Maheu, he began his career as a teacher. Yet his speech took on a very different tone. He began by reminding the audience:

> We are living in the era of *Homo sapiens*, from whom all human beings are descended and that science and technology should contribute, thus, to the greater welfare of man but not conduce to his annihilation or subservience. How can we silence our concern when we note that scientific research budgets for military purposes are continually growing in the majority of the industrialized countries. The amount spent today on military research in the six most highly developed countries exceeds 25 thousand million dollars annually and if we add to this figure these countries' arms expenditure, we reach the staggering figure of 250 thousand million dollars. To take the first figure, 25 thousand million dollars per annum to perfect the means of destroying man represents, by and large, the gross domestic product of all the African States invited to take part in this Conference.[16]

Senghor had served France in World War II—he spent two years in Nazi concentration camps after getting captured in 1940—so he knew a lot about the destruction of man. He joined the French Resistance on his release, and in 1946 he was elected on a socialist ticket to one of the two seats representing Senegal at the French National Assembly. He was reelected in 1951 and in 1956. Senghor was instrumental in fighting for Senegalese independence. He believed that small, fragmented West African states would be economically unviable. After the French government introduced self-government via the *loi cadre* in 1956, he worked to create a federation between French Equatorial Africa and French West Africa—the Mali Federation, in 1959. It was short-lived. Senegal and Mali split and became independent republics in August 1960, and Senghor was unanimously elected president.[17]

At CASTAFRICA, Senghor went on to detail the hardships many African nations faced during the successive droughts in the Sahel zone, and the "well-nigh insurmountable difficulties" they posed for their developing economies. Given the countries' largely agricultural economies, vulnerable to the vagaries of the climate, Senghor noted that it was "not surprising that our scientific and technological possibilities must be basically slanted towards the seeking of practical solutions."[18] Nevertheless, Senghor was a major proponent of black African creativity and innovation. With Aimé Césaire and Léon Damas, he developed a movement around Négritude—a celebration of black culture and arts akin to the Harlem Renaissance that took hold among black Francophone Caribbean and African intellectuals in Paris beginning in the 1930s—and wrote his own poetry in his spare time. Toward the end of his address, he made it clear that a major goal of his was for Senegal to develop its own scientific and technological research capacities so that it did not remain dependent on Western inventions.

The desire for more high-tech, modern solutions permeated the CASTAFRICA proceedings. It was this vision—of advanced African societies, more akin to a real-life Wakanda than to Arcadia—that drove a very different interpretation of what "appropriate" technology meant.[19] A survey of African scientists and policy makers was conducted for the conference on "technologically feasible futures for Africa." The final report argues as its frame of reference:

> It must first be borne in mind that Africa's economic and technical possibilities are limited for the time being; that the leeway it has to make up, though sometimes estimated on the basis of dubious criteria, is great; and that its needs are all urgent. From this some people infer that it should confine itself to making the most of past achievements, that is, to applying techniques which have proved satisfactory elsewhere. In their view, the developing countries ought not to serve as a testing-ground for daring new experiments which may or may not yield results. But another point of view was adopted on the occasion of the CASTAFRICA conference. This was that if a society confines itself perpetually to adapting what others invent, that society will perpetually lag behind the others. Consequently, if leeway is to be made up where it occurs, we must realistically and imaginatively set about building hypothetical and, in the main, utopian situations from which will perhaps emerge the revolutionary solutions alone capable of rapidly changing a pattern of relationships which is not for the moment in Africa's favour.[20]

The result was a list of forty-four priority technologies for transfer, in sectors such as transport (e.g., helium-filled airship, hovercraft, "all-terrain" vehicles, pipelines to transport mineral and agricultural products), communications (e.g., satellite communications, laser links), power (e.g., solar energy devices, desalination of seawater with nuclear energy), industry (e.g., pulp and paper plants, new materials and techniques for road surfacing), education (e.g., "automated teaching techniques"), and health (e.g., birth-control techniques, helicopter-transported operating theaters).[21] These imported technologies would, at first, help make up the technological gap for African societies. But the CASTAFRICA recommendations also recognized that many technologies that would be appropriate to an African context had not yet been invented—so they did not limit their definition of appropriate technology to those "old" technologies that existed already and would need to be adapted to work in rural or low-resource settings. Rather, the report expressed hope that investment would be made in the research and development of new appropriate technologies within Africa.[22]

CASTAFRICA emphasized African ingenuity, the training of African scientists, and the benefits of "vertical" technology transfer (the production and transfer of technologies within Africa) as opposed to "horizontal" technology transfer (the importation of technologies from developed countries). This fit broadly within the goals of the larger Group of 77 at the United Nations, the coalition of developing countries who made the call for the NIEO. The G-77 was formed officially in 1964 by the Joint Declaration at the UN Conference on Trade and Development (UNCTAD), although it built off earlier attempts at developing country solidarity as expressed at the Bandung conference of newly independent African and Asian states in 1955 and the Non-Aligned Movement, declared in Belgrade in 1961. Keeping production and technology transfer within Africa was also in line with visions of postcolonial federalism and self-determination advanced by many African leaders, including Kwame Nkrumah of Ghana.[23]

The CASTAFRICA interpretation of appropriate technology was, by Western standards, very technically sophisticated. It pushed back against the idea that simple technologies would be considered sufficient by postcolonial policy makers. Dissatisfaction with the Schumacherian version of "appropriate technology" was pronounced at the CASTAFRICA conference, as was a feeling that UNESCO was merely humoring the African ministers by holding the conference. While the United States and Soviet Union both sent observers to the meeting, only one of the eight major donors invited—the Ford Foundation—sent delegates to the meeting. This complicated negotiations for technology transfer, as the donors that could most readily fund many of the desired technologies were not present. The United States was a particular target for demands for these new technologies, both because of its perceived supremacy in science and technology relative to European donors and because its role in the Cold War—and the desire to keep many G-77 nations within its sphere of influence—gave many newly independent nations leverage in negotiations.[24]

UNESCO's sponsorship of CASTAFRICA, as well as a corollary conference focused on Asia (in 1982) and a series of smaller regional meetings, would eventually lead to charges that the organization was too political. This would come to a head in 1984, when the United States, under President Reagan, withdrew its membership from the UN organization. It was followed, a year later, by the United Kingdom under Prime Minister Margaret Thatcher. Both countries argued that UNESCO had "become a forum for Soviet and radical Third World initiatives."[25] The British minister for overseas development Timothy Raison asserted that the organization had become "harmfully politicized" and had "been used to attack those very values which it was designed to uphold."[26] Together, the countries contributed just under 30 percent of UNESCO's budget.

∴

Halfdan Mahler was famously sympathetic to the desires of the Third World and calls for a new international economic order.[27] He took office as director general of the WHO in 1973. His tenure was marked by sweeping changes he introduced to the organization, as he brought with him an activist focus on health equity.[28] Mahler was a strong, charismatic leader, whose influence over the organization meant he played a major role in policy and agenda setting for member states.[29] Established in the wake of World War II as a specialized UN agency for international health, the WHO was primarily concerned with disease eradication programs during its first two decades.[30] In the latter half of the 1970s, under Mahler's direction, a central concern at the WHO became meeting basic needs through the provision of primary health care. Appropriate technology was a key point.[31] Mahler was himself an appropriate technology evangelist, and he made it a frequent theme in his addresses to the organization in the mid-1970s.

The concern over "basic human needs" at the WHO echoed a similar priority that had been announced at the World Bank a year earlier, under the similarly strong and charismatic leadership of Robert McNamara. McNamara brought international health programs under the bank's mandate for the first time. Part of the Bretton Woods financial system established in 1944 under the leadership of Harry Dexter White of the US Treasury and Schumacher's former mentor John Maynard Keynes, the World Bank had to that point focused on relatively small, conservative loans for postwar reconstruction and international development through capital projects.[32] McNamara assumed the bank's leadership in 1968, appointed by President Johnson. He had a more holistic vision for what constituted development and an understanding that many large capital projects only benefited the relatively elite people who could, for example, access the power grid. In 1974, he shifted the bank's focus to meeting the "basic human needs" of all people, to alleviate poverty from the bottom up, which he thought would be more effective than waiting for the wealth generated through modernization projects to trickle down. This dovetailed with a wider shift in the discourse on human rights, from a focus on equality to a focus on sufficiency, or what Samuel Moyn has called "global subsistence rights"—the minimum necessary for survival.[33]

Appropriate technology was a key means by which the World Bank sought to achieve the goal of meeting basic human needs, although its interpretation focused primarily on the labor intensity of appropriate technology.[34] Closest to Schumacher's original goal of reducing "immiseration" through rural employment, the bank engaged in economic studies of whether labor could be substituted for capital in modern manufacturing. This was perhaps directly influenced by Schumacher himself, who had participated in an UNCTAD expert group on diversification and had favorably impressed the chairman, World Bank staffer S. R. Sen, to the point that he wrote to McNamara (attaching Schumacher's article "Intermediate Technology: The Missing Factor in Foreign Aid") requesting that he extend an invitation to Schumacher to deliver a lecture at the bank.[35] There is no evidence that McNamara followed through, although bank staff did make appropriate technology a key part of their development agenda.

In July 1976, the bank published its policy regarding the use of technology in international development. It held three primary tenets:

1) That the technology used in the projects [the World Bank] finances should be appropriate to development goals and to local conditions;
2) That the Bank, by itself or in collaboration with others, should promote innovations needed to make available to developing countries technology appropriate to their needs;
3) That Bank-financed projects should develop local capacity to plan for, select, design, implement, manage, and, when necessary, to adapt and develop appropriate technologies.[36]

The policy outlined a process for evaluating the appropriateness of a technology based on local conditions but held that all technologies should make "full use of abundant labor while minimizing the use of scarce capital and technical skills." Most of the examples cited in the policy document revolved around generic hand tools and small equipment for rural agriculture. In health care, however, the bank was quite specific in its funding goals:

> In the health sector, the Bank places emphasis on local identification of health needs, use of paramedical manpower, simplification of therapy and drug supplies and construction of low-cost structures by communities. The Bank discourages proposals for hospital-based treatment facilities, professionally staffed mobile clinics and advanced training of

> medical manpower. Instead, it promotes programs of vector (disease transmitting agent) control, health education and simple preventive and curative care, the utilization of community health auxiliaries (as contrasted to the highly skilled nurse or physician), and small-scale, community-based facilities such as dispensaries (as opposed to sophisticated hospitals).[37]

The contrast between what the bank did and did not see as appropriate for the developing world was quite stark.

Like his World Bank colleagues, Mahler was also prepared to be specific in delineating which technologies were "appropriate." In 1975, he called for the development of the "Essential Drugs List" for international health, which he thought would help to address issues of pharmaceutical overuse (in the First World) and inaccessibility (in the Third).[38] He viewed essential drugs as key appropriate technologies—while not labor-intensive, they were the right "scale" for most health systems, that is, not too new and sophisticated, but drugs that would just meet real needs—and he issued a more widespread appeal for careful consideration in the choice of health technology in his later report for 1975 to the Twenty-Ninth World Health Assembly. In the introduction, he said it was "unlikely that the build-up of centralized services based on advanced health technology will effectively meet the overwhelming day-to-day health requirements of the majority of people living in rural areas. We must remain conscious of how often the medical and health care solutions of the industrial world are neither practicable nor acceptable for developing countries, because of cost and inefficiency."[39] The overriding logic was one of scarcity and triage. He went on, "We must resist the temptation to introduce sophisticated medical practices such as intensive care units into less privileged parts of the world, where they are irrelevant when viewed against a background of much more urgent health problems."[40] In his keynote address in Geneva a year later, on May 3, 1977, Mahler again urged the member countries of the WHO to break the "chains of dependence on unproved, oversophisticated and overcostly health technology."[41] Rather, the focus should be on universal access to primary health care, including essential drugs, or what the WHO codified in 1977 as "health for all by the year 2000."[42]

Appropriate technology became one of the central tenets of primary health care as outlined at the WHO's International Conference on Primary Health Care, held in Alma-Ata, Kazakhstan, September 6–12, 1978. The conference is frequently invoked by historians of international health, as it signified a rare instance of unity and support for the type of broad-based health improvements that many developing countries sought.[43] Mahler foresaw some difficulties in its implementation, particularly with private industry: "We shall undoubtedly have to face many political problems. Some of them will derive from commercial and professional interests, where they are touched, for example, by the generation of appropriate technology for health, by the adoption of drug policies aimed at providing essential drugs for all and establishing drug industries in developing countries."[44]

At Alma-Ata, appropriate technology was heralded as a way forward but not well defined. During follow-up conferences aimed at implementing primary health care, the organization was ambivalent about the importance of conceptual definitions of appropriate technology. As an illustrative example, in a report on the Inter-Regional Workshop on Appropriate Technology for Health (ATH) held in New Delhi in March 1981, the chairman "observed that the concept of ATH addresses as its priority the most basic need of the poorest people. ATH, according to him, is a political and cultural phenomenon which attacks the current technological orthodoxy."[45]

During the workshop, there were two presentations on conceptual definitions. In the first, although "it was made clear that while it would be unwise to cloud the practicability of the issue by conceptual considerations," basic definitions of *technology* and the connotations of *appropriate* were given. The second elaborated on what appropriate technology for health was by focusing on how it interacted with the WHO's goal of "health for all by the year 2000." Here, the suggestion was that "ATH should be focused at all the vital interfaces in the implementation of [primary health care] activities . . . the focus of ATH is on review of the appropriateness of existing techniques and search for better tools." The report concluded somewhat tautologically, indicating that appropriate technology should focus on "operationalization" rather than conceptual definitions—"as [appropriate technology for health] is already understood in the context of various countries because appropriateness of technology depends also on a given situation." This concern over conceptual versus practical definitions was nonetheless prescient, as putting theoretical ideals into practice was exactly what Schumacher himself had struggled with when he founded the Intermediate Technology Development Group.[46]

∴

As the Vienna conference unfolded in late August, 1979, the years of preparation seemed to have come to nothing. The United States was condemned by countries in the NIEO for the failure of the conference before it even started. The Cameroonian scholar John W. Forje published a manuscript just before the conference kicked off, *The Rape of Africa at Vienna: African*

Participation in the 1979 United Nations Conference on Science and Technology for Development.[47] He dedicated the monograph "to Patrice Lumumba, Kwame Nkrumah, Amilcar Cabral, Eduardo Mondlane, Stevie Biko, Murtala Mohamed, Augustin N. Jua, Z. A. Abendong, S. A. George, Frantz Fanon, Nelson and Winnie Mandela, H. G. Fonyonga, Paulina Nagwa, Elias Mbanam, Lydia Leosongah, Jemica Lang, Gert and Alis Henriksen, and the future generation of Africa." It was a scathing critique of how, in his view, African countries had been sidelined and dismissed by the United States and its allies in the preparations for the Vienna conference—the conference was, in his words, a "fait accompli for Africa."[48]

Based on each country's precirculated position paper, Forje's analysis positioned American actions (and inaction) as a form of neocolonialism and questioned the utility of engaging in such multilateral forums in the future. In particular, Forje saw the machinations of global capitalism as violently exploiting developing countries. He wrote that the "trade preferences and the quasi code of technology transfer that will be handed to developing countries from the Vienna altar will be one of unequal industrial cooperation, i.e., industrial cooperation with firms that are on the verge of liquidation will be resurrected with new blood, cheap labor and markets in the developing countries."[49] The reluctance of "metropolitan" country governments to curb the activities of multinational corporations, he went on, explained why the apartheid systems in South Africa and Rhodesia had survived trade boycotts and economic sanctions. Although Forje admitted that many capital-intensive Western technologies were not well suited to the needs of developing countries (citing fishing trawlers in Ghana displacing the traditional fishing industry and causing mass unemployment, among other examples), he described appropriate technology as an "old trend," written into developing countries' position papers by the ruling elites who had entered into an "unholy marriage" with Western elites to preserve the status quo.[50]

Throughout the conference, a group of mostly anonymous journalists (rumored to be from Scandinavia) published a daily conference newsletter, *Retort,* which was freely available to delegates as they walked in the main doors of the Stadthalle each morning (after a minor incident on the second day, in which all copies were initially seized by the security guards "until clearance came from a higher authority").[51] Often reporting on the more informal interactions at the conference, *Retort* did not shy away from controversy or from taking political stands—rather, its authors indulged in attention-grabbing headlines that made their position on the various issues clear. They described Father Hesburgh's opening speech as one that "fluctuated rapidly between rhetorical flourish and pragmatic propositions," which, together with remarks he conveyed on behalf of President Carter, centered on the need to consider what was possible within the strained economic climate.[52] "All is not well in the international market of technology," Hesburgh relayed. "Transferred technology is often inappropriate to local needs, as well as wasteful, and insensitive to environmental impact. Such transfers are bad business."[53] In characterizing most transferred technology as "inappropriate," he was in effect saying that the US government would not supply the developing world with the modern technology that its leaders wanted.

By day 4, *Retort* was already proclaiming the conference's failure. With a full, front-page spread of white lettering printed on a black background, the headline declared "STOP THE CONFERENCE!—we want to get off."[54] Although only three full days had elapsed on the Vienna conference program, the authors wrote that "already some of the worst fears of observers—that commitment to change professed in bold conference declarations may be worth little more than the government notepaper they are written on—are rapidly coming to fruition."[55] Their main concern was that it already seemed at that point that there was no chance that the developed nations would support the central demand of the G-77, namely the creation of a fund to support science and technology transfer to G-77 countries that would total no less than $400 million by 1980, $600 million by 1981, and $4 billion by 1990.[56] Although Father Hesburgh had agreed to the creation of the fund in principle, he said that there would be issues with Congress on assigning it any money. He also told Mahmoud Mestiri, the Tunisian special representative to the United Nations who led the G-77 in negotiations (as Tunisia was the presiding country of the G-77 that year), that, although the United States was supportive of enhancing developing countries' technological capabilities and "growing agreement" on sharing information systems, he "did not think that they were likely to achieve full agreement on transfer of technology, the role of multinational corporations, or patents."[57] Rather, these issues would have to be sorted out in concert with other bodies, including UNCTAD on technology transfer and the Paris Committee on patents. Mestiri viewed this as punting the issues further down the line, and the G-77 grew discouraged with the meeting overall.

As Martin Kaplan, a consultant to the WHO for the conference wrote afterward, "there was minimal progress in facilitating access to industrial information, and to patent rights and transfer of technology in general. A global information system was agreed in principle, but its structure and character were left undefined."[58] Ambassador Wilkowski, who would go on to chair the board of VITA, described the conference as getting bogged down in minor details, writing in her memoir, "I even feared I might be having a heart attack in exasperation over some minor language differences: was it to be an 'a' or a 'the' in a particular sentence?"[59] The NIEO did not achieve the goal of securing greater access to advanced technology through technology transfer. Instead, the United States and other Western governments emphasized the development and use of simple appropriate technologies.

Funding for the development of new appropriate technologies did not come easily, either. The Institute for Scientific and Technological Cooperation (ISTC) that President Carter's envoy, Dr. Gilbert Omenn, proffered at the Institute of Medicine conference proved a significant sticking point with Congress in the final months of preparations for the Vienna conference. It was intended to be the administration's flagship offering at the conference—a research and development enterprise for appropriate technology solutions to developing country problems.[60] Modeled on Canada's International Development Research Centre, it would fall directly under the umbrella of the International Development Cooperation Agency (IDCA), which at the time also oversaw USAID within the Department of State.

Congressional opposition focused mainly on the potential expense of the new institute as well as the view that it duplicated functions that already fell under USAID's purview. Initially rejecting the bill, the Senate eventually passed the International Development Cooperation Act of August 14, 1979 (Public Law No. 96-53), just six days before the start of UNCSTD, which established the institute but did not appropriate a budget for it. President Carter issued an executive order establishing the institute (again) that September, but again was unable to assign it any money.[61] On February 5, 1980, IDCA director Tom Ehrlich testified before the House Committee on Foreign Affairs, pleading the case for an ISTC budget of $95 million for fiscal year 1981, of which $57 million was to be transferred over from USAID's budget for the continuation of its research and development programs.[62]

The debate also played out on the editorial pages of the *Washington Post*.[63] Democratic senator Dennis DeConcini of Arizona was one of the more vocal opponents, arguing that after USAID's own $275 million annual research and technical assistance budget, the ISTC was "simply an unnecessary addition to an already top-heavy, overgraded foreign assistance bureaucracy. Indeed, it is nothing short of a bureaucratic bonanza."[64] As a large part of the ISTC research agenda was drug and vaccine development, in consultation with US academics and industry, Werner Fornos, the director of the Population Action Council, wrote the ISTC off as "just another research subsidy," and posed the question: "Should foreign aid be spent on research in laboratories at home, or on effective programs within the countries it is intended to help?"[65]

The ISTC never did get funded. It fell victim, ultimately, to the massive budget cuts of fiscal year 1980.[66] Instead, Congress eventually approved the $12 million Program on Scientific and Technological Cooperation under USAID. The USAID administrator Douglas Bennet, meanwhile, began to recruit for a science and technology adviser, in consultation with IDCA director Tom Ehrlich; Frank Press, director of the Office of Science and Technology Policy at the White House; and Thomas Pickering, director of the State Department's Bureau of Oceans and International Environmental and Scientific Affairs. Staffing that post was difficult, too. Donald A. Henderson, of smallpox eradication fame, turned down the post, as did the chemist Cyril Ponnamperuma. The job eventually went to a rear admiral in the US Public Health Service.

∵

The Vienna conference, like its predecessor, UNCAST, became another episode in which the United States failed to live up to developing countries' expectations. While appropriate technology emerged as a rallying point and solution for many different voting blocs, the distortions in how *appropriate* was interpreted meant that this semblance of alignment fractured when it came to defining practice-oriented policy measures. For the US delegation, appropriate technology did a lot of work—supported by economists and multinational organizations like the World Bank and WHO, it seemed both morally and economically justified; focused on simple, off-patent technologies, it was cheaper for USAID to implement and consistent with US private-sector involvement in foreign aid; framed as the most scientifically and technologically efficient means of development, the US delegates could claim to be in agreement with NIEO demands for the best of American technical and scientific know-how. In other words, appropriate technology could be used to placate many different constituencies simultaneously, aided by the malleability of its definition.

Between competing visions of appropriate technology, it is not surprising that the American one won out in Vienna. As a major donor and purveyor of high technology, although the United States allowed and even encouraged open-ended interpretations of appropriate technology in its rhetoric, the delegation worked hard to maintain its control over which version of "appropriate technology" was put into practice.[67] There was therefore a hierarchy of authority in defining what was appropriate, and "official" venues for its elucidation were often undercut by subversive efforts on behalf of the Third World, including the "shadow conference" at UNCAST and the special ad hoc meeting at CASTAFRICA.

There were many costs to this restricted definition of appropriate technology. As the village or community became the unit of analysis for appropriate technology programs, remote areas became sites of experimentation for the distribution and uptake of new technologies—as was proposed in Mauritania.[68] While donors foreclosed access to some technologies on the grounds they were not "appropriate" for developing nations, technologies that were deemed appropriate were thought of as universally applicable—dispersed with little consideration for local context or the actual needs of the supposed beneficiaries.[69]

In retrospect, the most successful part of the Vienna conference may well have been the simultaneous NGO Forum, which was organized as an appendage to the official conference after the nonstate actors protested their exclusion from the program.[70] Operating outside of traditional state-based bilateral and multilateral institutions, nongovernmental organizations (NGOs) received a substantial boost in their numbers and funding with the cutbacks to the American foreign aid budget. As the desire for "small government" took hold, many functions that USAID had formerly had the in-house technical expertise to perform were outsourced to the multitude of organizations that sprang up to fill the void. The ascendance of NGOs therefore marks the ascendance of a form of "shadow" governance—one outside the direct control of the US citizenry—but one that the government is reliant on to fill vital roles. As we will see, this degree of independence afforded to the nongovernmental sector in international health would soon allow for even greater differentiation in the interpretation of appropriate technology. These new organizations, focused on solving development and health problems through novel science and technology, pushed the boundaries of what Western governments thought was "appropriate" for the developing world.

[CHAPTER SIX]

The Silver Bullet Boys

> The advance of technology is based on making it fit in so that you don't really even notice it, so it's part of everyday life.
>
> Bill Gates[1]

> Once a new technology rolls over you, if you're not part of the steamroller, you're part of the road.
>
> Stewart Brand[2]

Dr. Richard Mahoney clearly remembered a memo that appeared on his desk at the Ford Foundation's New York headquarters in 1973. Titled "Contraceptive Introduction, Manufacture, and Supply," it was written by the Canadian obstetrician-gynecologist Dr. Gordon Perkin, then in the employ of the foundation at its Brazil field office.[3] In the memo, Perkin argued that for family planning programs to reach their full potential, "the supply of contraceptives needed to be dramatically increased."[4] He made the case that access to high-quality, affordable contraceptives would dramatically improve health outcomes in the developing world. "I picked up the phone and called Brazil," Mahoney remembered, "even though in those days it cost US$100 every five minutes to make an international call. And I said, 'Gordon, let's do it. I'm ready. Whatever you need, I'll do it.'"[5]

Four years later, Perkin left his job at the Ford Foundation to officially launch a new nonprofit, known in its infancy as the Program for the Introduction and Adaptation of Contraceptive Technology (PIACT), headed by the contraceptive expert Dr. Gordon Duncan. "Any idea was accepted, and any idea could be tried," said Mahoney. "I am often reminded of that culture when I read about the culture of the tech startups in Silicon Valley. How people were just totally devoted to trying to make something work. You spend hours and hours thinking about what to do, and it's a seven-day-a-week enterprise, and there's lots of fun and laughter. We were great optimists, all of us, and we just felt that this would work. That family planning had to work."[6]

The description is an encapsulation of many important trends in 1970s international health and foreign aid. Privatization and quantification were major commitments at the time, well before the "neoliberal" 1980s. PIACT, later known as the Program for Appropriate Technology in Health (PATH), benefited from the push to contract out foreign assistance to nongovernmental organizations (NGOs) after USAID faced major budget cuts in the 1970s—the rise of NGOs and the subsequent expansion of their footprint in the developing world via field offices created an alternate infrastructure of development aid.[7] Today NGOs are typically understood to be nonprofits; however, in the late 1970s as they first proliferated in the development sphere, they were known as PVOs—private and voluntary organizations, a category that merged for- and nonprofit outfits, in part because there was a simultaneous push to more heavily involve the private sector in development aid to help make up for budget shortfalls at the time.[8] This outsourcing of development, or "projectification"—dividing the gargantuan task of "development" into project-sized chunks that could be contracted out for two to five years at a time—also led to a focus on metrics enabled by the distribution of small-scale technologies, such as number of doses administered, number of devices delivered, or percentage increases in infant growth curves.[9] NGOs were able to fund projects with contributions from multiple donors, thereby gaining the ability to set their own programmatic priorities. PATH set its sights on developing appropriate technologies for international health and, in the process, transformed notions of what appropriate technology was.

Although not widely known outside of the field of global health, PATH has played an instrumental role in shaping the form and function of the field as we know it today. Its vision of appropriate technology—one of novel technologies designed specifically for the exigencies of the developing world—contrasted with the vision the US delegation had held to in Vienna. The founders' start-up zeal paralleled the rise of other tech firms based up and down the West Coast of the United States, from Seattle to Silicon Valley. Formed at a moment when appropriate technology was considered a potential solution to many of the world's problems—aided by popular media like the *Whole Earth Catalog*—PATH leveraged the concept to promote its technological fixes.[10] Not least of these problems was overpopulation.

Concerns about planetary overpopulation began to grow in the late 1960s, pushed along by the burgeoning environmentalism movement and the 1968 neo-Malthusian publication by biologist Paul Ehrlich, *The Population Bomb*.[11] Which populations were supposedly overreproducing, posing a threat to the carrying capacity of Earth, was just as much a concern—the black and brown populations in the Global South were the locus for most of the population-bomb fears.[12] Development programs in the 1970s therefore turned to family planning as a means of decreasing birth rates.[13] Empowering women through education became a secondary goal, as studies showed that higher maternal education levels corresponded with lower fertility rates.[14] Fewer children would also mean a reduced likelihood of maternal deaths and, at least in theory, would enable greater women's equality.

As a means of coordinating these goals, in 1975, the United Nations declared the following ten years to be the "Decade for Women." It kicked off at the first UN women's conference, held in Mexico City. A midpoint conference was held in Copenhagen in 1980, and a final conference at the end of the decade was held in Nairobi in 1985. In addition to access to education and family planning, violence against women, landholding rights, and other basic human rights were part of the larger agenda to achieve women's equality.[15]

The UN Decade for Women aligned closely with the peak years of the appropriate technology concept's popularity. This was fitting, as many technologies designed to be "appropriate" were inherently very gendered, and the populations who were subject to the most concern about overpopulation in the developing world were the same ones Schumacher feared would leave their rural subsistence lives without better tools. The appropriate technology concept received the most attention in the health, agriculture, and water, sanitation, and hygiene (known colloquially as "WASH") sectors of development. What these have in common, other than an orientation toward health and nutrition, is that they are most often the responsibilities of women in much of the world. The water pumps, grain-husking machines, and infant growth-monitoring devices that have been distributed through foreign aid programs have been largely designed and selected by white men in the Global North for use on or by women of color in the Global South. The job of maintaining these technologies also usually falls to women, although that labor is rarely visible. That appropriate technology was, by the classic Schumacherian definition, labor-intensive meant that women were disproportionately doing that intensive labor.[16]

Within Schumacherian visions of the appropriate technology concept, it was relatively easy to construct a binary of Western male innovator and designer versus non-Western female maintainer and user. This is a clear oversimplification. Yet there was very little discussion, among US foreign aid policy makers or proponents of appropriate technology in the developing world of the 1970s, of how appropriate technology affected women specifically. Rather, they touted the concept as benefiting the nation and furthering the development of communities as a whole. These grand claims, made almost exclusively by men, concealed the true users of the technologies—women. If appropriate technologies did aid development, they did not do so sitting idly or of their own volition, as in Gandhi's depiction of the ideal village. The labor of the women using the technologies was the force driving that development.

The delegates to the final UN Women's Decade conference, in Nairobi in 1985, were cognizant of these inequities. In the "Forward Looking Strategies for the Advancement of Women to the Year 2000," a section on science and technology read that "governments should reassess their technological capabilities and monitor current processes of change so as to anticipate and ameliorate any adverse impact on women, particularly adverse effects upon the quality of jobs."[17] It also recommended increased women's participation in science and technology decision-making, increased funding for women's education in scientific and technological fields, and "intensified . . . efforts in the design and delivery of appropriate technology to women." This latter specified that "the implications of advances in medical technology for women" should be studied with particular care.[18] In a long section on health care, the conference reaffirmed the centrality of birth control in the fulfillment of women's other rights.[19]

If the orthodox appropriate technologies were implicitly gendered, PATH and its predecessor organization created appropriate technologies focused on family planning that were very explicitly gendered. Its position as an outsider to the US government enabled the mission. Even under the relatively progressive administration of President Jimmy Carter, the United States has historically been leery of funding family planning programs directly due to their still-controversial status within some religious constituencies in the country.[20] This chapter argues that NGOs were able to reinterpret appropriate technology according to their own prerogatives, particularly in women's health programs. In doing so, they pushed what was "appropriate" away from the simple, off-patent technologies of old and toward a model of rendering advanced drugs, vaccines, and devices radically affordable for the women and children of the Global South.

At the same time, we can see how the rise of PATH and other NGOs promoting international health in the late 1970s set the groundwork for the massive scale of private donor-driven global health programs in the twenty-first century. Multiple-donor projects led by NGOs diluted the agenda-setting power of state-based bilateral donors, while the pioneering of what are now called public-private partnerships has vastly increased the role of the private sector in global health projects. While these structural changes eventually led to the consolidation of global health investment led by a few, behemoth private organizations, they initially led to the decentralization of project management both in the United States and in recipient nations.

Privatization was therefore not initially the trickle-down effect of a neoliberal ideology intentionally adopted by the US government, but ironically it was a way to make ends meet for projects that had both limited funding and a need to create technologies that were affordable for recipients.

∵

In 1973, as Bill Gates and his soon-to-be Microsoft cofounder Paul Allen were tinkering away with computers in Seattle and tech companies and the venture capital firms they relied on were gearing up in Silicon Valley, another set of cofounders hatched their idea for a startup. The trio had very different, though complementary backgrounds. Dr. Gordon Perkin was raised in Toronto, and after receiving his MD from the University of Toronto in 1959, he worked as a family doctor before his work for the Ford Foundation overseas. He brought knowledge of developing countries' needs and direct field experience to the group. Dr. Gordon Duncan had a career in academia before he became the vice president of research at the Upjohn pharmaceutical company. Intimately familiar with the private sector, he was the developer of the vaginal ring contraceptive and was also instrumental in getting US Food and Drug Administration (FDA) approval for both the emergency contraceptive pill and Depo-Provera—the latter of which took twenty-eight years—and he brought this knowledge of technology development and regulation to PATH.[21] Dr. Richard Mahoney, a PhD in chemistry from University of California, San Diego, brought expertise in project management. After a stint working on Bobby Kennedy's presidential campaign, Mahoney directed his idealism toward the Ford Foundation's Population Office in the early 1970s and became the foundation's youngest program officer, in charge of contraceptive development, shortly thereafter.

Mahoney and Perkin began working on PIACT in 1973 as a side project. The organization was founded officially in 1976, and Duncan became its first chief executive officer, while they worked out of apartments colocated on Batelle's campus in Seattle (carpeted in orange shag, according to legend). Perkin took over as chief executive officer two years later, and Mahoney quit Ford officially at the end of 1978 to join the other two full-time, moving to the Philippines to become the regional representative of PIACT in Asia. Both Perkin and Mahoney left Ford on very good terms. Their former boss, the chief program officer Dr. Oscar Harkavy—known to friends and close colleagues as "Bud"—helped ensure that the Ford Foundation became PIACT's first funder and one of its most consistent early boosters. With a pledge of $92,000 in seed money from the foundation and the donated office space, the three cofounders set to work designing novel contraceptive programs for the developing world.[22]

What made their approach novel, at a time when concern over the "population bomb" and family planning programs was at its peak, was their attention to the specific needs of local communities at the outset, rather than advancing the technology first and then finding an application for it post hoc. Early on in the development process for a new drug or device, PIACT would engage with local stakeholders and—crucially—also find a commercial manufacturing partner, so that the end users and manufacturers were both able to shape the technology as it unfolded. This bottom-up model of development resonated with proponents of the appropriate technology movement, and in 1980, the organization was rechristened the Program for Appropriate Technology in Health (PATH).

What counted as a local community, and who could speak on that community's behalf, has been one consistent register on which PATH and development programs more broadly have been criticized. From the perspective of some stakeholders, an approach may be completely reasonable; from the perspective of another, it may appear senseless, discriminatory, or harmful, in spite of the project designer's apparently good intentions. As suggested by the twenty-eight-year FDA approval time, Depo-Provera, which Gordon Duncan worked on before he joined PATH, was one such controversial project. Although persistent safety concerns with the contraceptive method based on animal studies meant that Upjohn could not gain FDA approval for use on women in the United States (and thus the drug could not be legally distributed through American foreign aid), between 1972 and 1978 it was tested off-label on women at the Grady Memorial Hospital in Atlanta, which served primarily low-income women of color, and in the early 1980s on mentally impaired women under the care of the US Indian Health Service.[23]

There is no evidence that Gordon Duncan, who passed away in 2016, had any knowledge of or part in these trials. In interviews he recorded before his death, he seemed proud of his accomplishments in the field of contraceptive development—viewing them as liberating for women, and in countries with high maternal mortality rates, as life-saving interventions for mothers.[24] However, the Depo-Provera trials are part of a broader legacy both of testing obstetric and gynecological innovations on low-income or enslaved women of color—its own form of gendered and racialized labor—and of using marginalized communities in the United States as analogs to their counterparts in the developing world.[25] This form of "model village" was seen as generalizable across vast geographies. Indeed, PATH would sometimes trial products in the United States before using them in overseas development programs as a means of cost-savings—for example, workshopping health communications pamphlets in low-income communities of color in Baltimore and Alaska Native communities before adapting them for an overseas market.[26]

In its attempt both to innovate and to keep things cheap, PATH did run counter to most Western governments' prevailing notion of appropriate technology, which, while also low-cost, was decidedly less innovative. Far from PATH's heat-sensitive vaccine labels and rapid-diagnosis hepatitis B test kits, the government's vision was largely one of bamboo tube wells and wheelchairs made from repurposed bicycle parts. As Richard Mahoney explained:

> We made clear in our early proposals and our documents that when we referred to appropriate technology, we meant to take the best technology in the world and make it appropriate for people in developing countries, not take some idea about a household stove and make it out of old oil cans. I mean, we wanted the best technology possible and to make it so it's usable and affordable in developing countries . . . I think that was one of the things that made us attractive to donors. I think if we would have gone to them and said, "We're going to Mexico to make contraceptive steroids from the plants with local farmers," we would have been laughed out of the room. Whereas, we said no, we want to take the best contraceptives in the world and create ways of delivering them that will be appropriate and affordable.[27]

The trade-off for innovation was that PATH's technology was typically not locally sourced, as the classic Schumacherian definition of appropriate technology held. Rather, it can be read as what medical anthropologist Peter Redfield called a "second-best world" fix, in which innovations of the market supplant the traditional care offered by a functioning state or health system.[28] The ideal world, in which everyone had timely and affordable access to a modern, functional, well-stocked clinic, was a far cry from reality. PATH's response to development need with what Redfield called ethical design was never in the pursuit of profits but rather was animated by the conviction that there had to be a way to deliver modern medicine with minimal infrastructure or clinical support. They made do.

PATH's focus on local affordability—with many interventions designed to cost less than a dollar per patient, a threshold beyond which many innovations would be inaccessible to those who most needed them—meant that the margins they could offer manufacturing partners were often below the level that would interest major international pharmaceutical companies. For many of its early projects, therefore, PATH turned to lesser-known firms in countries like China and India that were more willing and able to work within extremely tight profit margins. In the early 1980s, the Vietnamese government asked PATH to help establish condom-manufacturing facilities. Peggy Morrow, a former vice president of PATH, recalled that at the time the major suppliers of condom-manufacturing technology were in the West and were both financially and logistically out of reach for the Vietnamese. PATH reviewed good manufacturing practices for condoms in different countries and, after visiting condom manufacturing firms in India, found that they were able to meet international standards. PATH then arranged a technology-transfer program between India and Vietnam with UN funding, with Indian companies providing affordable technology and training in manufacturing practice to the Vietnamese—an early example of South-South technology transfer.[29]

PATH's embrace of what would now be called public-private partnerships grew out of necessity and a quest for cost-effectiveness rather than any sort of ideological commitment to private-sector capitalism or free markets. The idea that private companies were friends not foes was also specific to the thinking of its founders and to Bud Harkavy, their initial booster at the Ford Foundation. Harkavy had encouraged Mahoney, while he was still at the Ford Foundation, to look into ways that the private sector's success in contraceptive research could be made more accessible for the needs of developing countries. Gordon Duncan had worked extensively in the private sector at Upjohn and Batelle so knew the ins and outs of approaching pharmaceutical company partners. Gordon Perkin, meanwhile, had written what came to be known as the "CIMS Memo"—"Contraceptive Introduction, Manufacture, and Supply," which had launched the effort. He argued that they needed the private sector to reach as many people as possible. "Collaboration with the private sector was something we focused on from the very beginning. We didn't see them as enemies," Mahoney recalled. "It was common in the early 1970s to think of the pharmaceutical industry as this rapacious, terrible enterprise where all they do is scheme of how to make profits and keep valuable products from poor people. This view didn't match with my experience at all."[30] During his time at the Ford Foundation, Mahoney had found that many pharmaceutical companies that he visited were very willing to work with them to make their products more accessible to the developing world. "I found people who were immensely devoted to the idea of family planning for people in developing countries, for poor people," he remembered. "They were working night and day to find ways to make the products available. Of course, they had to meet their investor's expectations also. They had to do a balancing act."[31] Mahoney extended those working relationships he'd developed at the Ford Foundation to PATH.

One of the earliest and most successful PATH projects was the vaccine vial monitor (VVM), a heat-sensitive label affixed to vials of vaccine solution that indicates when the vial has been exposed to too much heat to be effective. As with most of PATH's innovations, the idea sprang out of a desire to find solutions to common, real-world problems. Gordon Perkin recalled that the VVM was

> a very serendipitous development. I happened to read in *Popular Science* about this chemical and allied chemical they had that changed color with heat exposure. And so I contacted them and wanted to find out how close a match was between the kinetics of vaccines and the chemical that they had. And it turned out it was pretty close. That was really PATH's first big project and it was kind of an "Aha!" moment because we were trying to describe what PATH was to people who sort of had a blank look in their eyes. But when you describe the vaccine vial monitor and what it could do and how much wasted vaccine it could save, they suddenly got it and "Aha!" . . . It's really a wonderful example of appropriate technology because it was a sophisticated polymer technology that could be applied to a public health problem and address it, bringing about a solution and thus saving billions of dollars because they used to throw out as much as 20 to 30 percent of vaccines.[32]

The VVM has been required on all vaccines distributed by UNICEF or prequalified by the WHO since the early 1990s—today, more than 5 billion have been distributed.[33]

Adoption of the VVM was not without critics, however. In particular, officials at the Pan American Health Organization, the regional body of the WHO responsible for North America and South America, initially raised strong objections to VVM labels. Perkin recalled that PAHO had spent thousands of dollars training health workers throughout Latin America to throw out leftover vaccine vials or those that had sat out: "They'd open the 10-dose vial and only have two kids show up at the clinic for immunization, and they'd throw out the rest. So in some cases, the majority of the vaccine was wasted or thrown out." One PAHO official told Perkin, "[The VVMs] will come into Latin America over my dead body . . . we spent thousands of dollars training workers: if it sits out, throw it out. And you come along and send [VVMs] and show them we're liars."[34] To Perkin, the dispute was an important lesson—"every new technology threatens someone in the status quo and it needs to be addressed sooner rather than later." Another cause of tension between PATH and the WHO in the 1980s was PATH's close relationship to the private sector. "For many years, we, at PATH, were looked at with suspicion, for example, by the WHO . . . In some cases, we were not allowed to attend certain meetings at WHO because we were seen to be on the side of the private sector," Mahoney recollected.[35]

Relations between PATH and USAID were similarly confused. USAID had begun contracting work out to nongovernmental organizations (at the time, known as private and voluntary organizations) in the 1970s, due mainly to budget cuts.[36] PATH was an early contracting outfit. Longtime USAID staffers initially saw PATH, in some cases, as a "threat." "They didn't know what to do with us," Mahoney remembered. "They didn't know whether we were a private-sector company for which they had rules and regulations about whether they could talk to us on the phone or meet with us, or whether we were the public sector where we were friends and colleagues."[37]

This came to a head over one project PATH launched which aimed to implement a quality assurance protocol for condoms distributed through US foreign aid. According to Mahoney, it was well known that many of the condoms shipped to developing countries were faulty or became so under storage conditions in tropical regions. Akin to the VVM, he recalled: "We wanted to develop a better condom testing system to detect what lots were good and what lots were bad. And you could monitor them over time." He continued, "USAID saw our efforts as a threat because they thought what we were really trying to do was show that AID was buying shoddy condoms and shipping them to developing countries."[38] The dispute was resolved when the private sector embraced the standards that PATH had developed, and USAID, in turn, jumped on board.

Although public-private partnerships are almost necessary to secure funding in global health programs today, in their early decades many public-sector agencies—whether government based, like USAID, or multilateral organizations, like the WHO, questioned whether the private sector could work in good faith toward affordable health technology. Would the profit motive not inflate prices? Would intellectual property protections not create monopoly-like conditions around new research and development? PATH worked around these concerns both by focusing on appropriate technology, designed specifically to be simple and low cost, and by partnering with manufacturers in India and China that were more willing to work within extreme cost constraints. For Perkin: "By being appropriate, you weren't imposing these technologies on people who didn't want it. It was optional. You accepted it if you wanted to. And it helped address the problem. It wasn't . . . imperialistic, I guess." The economist E.F. Schumacher's book *Small Is Beautiful* was an influence, Perkin said, but PATH departed from his work in its focus on developing new, more advanced technologies. Skeptics, he said, used to refer to them as the "Silver Bullet Boys," assuming that their focus on technology was an idealistic search for the sort of magic bullets that German physician Paul Ehrlich and his contemporaries in the late nineteenth and early twentieth centuries searched for.[39] But their idealism broke ground that today is well worn. "Today you can't have a meeting at WHO concerning contraceptives or vaccines or other products without the private sector," Rich Mahoney noted. "I believe we had a good sense of where the future had to go."[40]

PATH's cofounders were wary, however, of developing technology in a vacuum, without the input of members of the communities where the technologies were meant to be used. The original board of directors consisted of experts exclusively from developing countries—it remained so for many years, in part because Ford Foundation funding required reports on the demographic makeup of grantee organizations' leadership. "That was one of our principles," Mahoney recalled, "We could not say we're for the developing world, but have a board of directors made up of Americans. It just didn't seem right. From the very beginning, we said we had to have a leadership from the developing world. And we did." PATH had a similar philosophy when it came to fieldwork. Mahoney spent a lot of time working in the Philippines, Thailand, and Bangladesh, and Vice President Peggy Morrow spent a considerable amount of time in China. "We said, we can't sit here in Seattle and try to solve the world's problems," Mahoney remembered, "We have to be in the field . . . I guess I can use the word 'visionary' in the sense that it has become a mantra that if you say you're working for developing countries, then you've got to be there."[41]

This orientation toward fieldwork eventually led PATH to open up field offices for project staff. In the early 1980s, this was relatively uncommon. A nongovernmental organization might go into a developing country and hire local people to carry out clinical trials, but it was rare to establish a lasting presence. The Population Council, an NGO focused on family planning, was one exception; however, its field offices were for the organization's social science rather than biotechnology work, which was done in New York and European laboratories.[42] PATH's first field office, established in 1979 by Mahoney in Manila, became a hub for assessing regional needs in contraceptive technology. At the same time, other relatively new NGOs were asserting their positions in major international forums, including at the UN Conference on Science and Technology for Development. As NGOs gained a solid foothold in development programs in the late 1970s, their embrace of appropriate technology was reflected throughout the discourse on agenda setting.

∵

After protesting their exclusion from the official UN conference program, the NGO Forum on Science and Technology for Development opened concurrently, on August 19, 1979, just a few blocks south in the Kongresshaus Wien. With symposia, panel presentations, and workshops all programmed, the NGO Forum—"open to all individuals and organizations"—was "aimed at providing a truly unique contribution to the deliberations."[43] While the "official conference" was focused on working out the "political mechanisms" for spurring science and technology in international development, the "alternative conference" considered "specific topics for action" by NGOs, including plenary sessions on nuclear energy, environmental change, and the role of multinational corporations.[44] The organizing committee, which comprised fourteen international NGOs and was chaired by Dr. Karim Ahmed of the Natural Resources Defense Council in New York, was particularly committed to working out a system of "information and financial mechanisms" to support NGOs doing work to implement programs post-UNCSTD. "One particular goal," according to Ward Morehouse, another member of the organizing committee, was "to see if we can create mechanisms for monitoring what happens after UNCSTD, and exposing it to public view in an effort to make governments keep up to their promises."[45]

The NGO Forum organizing committee also commented on the official proposals and position papers of the delegations to Vienna. They pointed out areas "such as the need for appropriate technologies, or the role of women in development," which the committee felt were underrepresented in the official program.[46] Dr. Ahmed and the organizing committee, like their UNCSTD counterparts, prepared for months in advance of the conference opening. Ahmed hosted a symposium in January 1979, in response to the release of the draft outline of the UNCSTD program of action, to discuss "technological alternatives for Third World countries."[47] "A major role of the NGO community stems from the ability to take independent stances on controversial issues," Dr. Ahmed wrote. "This freedom enables the member-organizations to raise questions critical to the international agenda and discuss them openly."[48]

The organizing committee formed eight task forces to review Vienna draft programs of actions. The task force's themes reveal the committee's priorities: health, nutrition, and population; culture and development; environment and energy; women's role; industrialization; Third World cooperation; communications; and food and agriculture. Rather than focusing on defining science and technology and debating how to obtain it, the NGO Forum was concerned with the social impacts that science and technology could bring.

The forum began with an opening plenary that featured an international panel of speakers, including Marxist economist Samir Amin and Russell Peterson, then president of the Audubon Society. In the symposia over the following days, other speakers included Katherine Elliott, a pediatrician originally of the ITDG who left the organization to found the Appropriate Health Research and Technologies Action Group in London, and Mansur Hoda, a close associate of Schumacher's who directed the Appropriate Technology Development Association in Lucknow, India. Appropriate technology pervaded the program, in everything from the health sector to the creation of model communities.

One panel, "Alternative Lifestyles, Community Participation and Development," focused on the creation of intentional communities. Reminiscent of the model villages that early communitarians advocated for, these communities took different forms. Stephen Gaskin, a counterculture spiritual leader, spoke about the commune he and his wife, Ina May Gaskin, founded in 1972 in Summertown, Tennessee, called the Farm. The Farm was (and remains) home to several appropriate technology initiatives: an eco-village demonstration center complete with solar panels and adobe housing; a thriving, self-taught midwifery practice that serves not only the women of the commune but also the many visitors who came to give birth at the birth center and the nearby Amish community; an independent book publisher, the Book Publishing Company; and, as a completely vegetarian community, several innovative initiatives in mushroom growing and soy-food processing.[49] Gaskin also started the humanitarian organization Plenty International in 1974. Plenty engaged in humanitarian service projects, centered on appropriate technology in Central America, South Asia, and Africa as well as in underserved communities in the United States, including American Indian reservations and the South Bronx.

Also on the panel, Dr. Devendra Kumar, director of the Centre of Science for Villages in Wardha, India, was directly inspired by the Gandhian village movement. Kumar began the center in 1976 on the premises of Maganwadi, where Gandhi had started the All India Village Industries Association in 1934. Kumar's economic theories were, like those of Gandhi and Schumacher, informed by J. C. Kumarappa, and he began the Centre of Science for Villages to "act as a center for transfer of technology and be a bridge between the portals of National Laboratories and doors of the Rural Mud Huts."[50] The center showcased and sold technologies including agro-waste stoves, beekeeping equipment, bull-drawn carts, and organic farming implements.

The appropriate technology community that Gaskin and Kumar's co-panelist, Dr. William "Bill" Ellis, founded and managed, was far less tangible but in many ways had further reach. An early iteration of what he called a "network," the Transnational Network for Appropriate/Alternative Technology, known as *TRANET*, was founded in 1978. It was in many ways similar to the work of VITA, the ITDG, *Whole Earth*, and other "catalog" producers in the later 1960s and early 1970s, and it was featured in *Small Is Possible*.[51] A 1982 article in the *Christian Science Monitor* positioned TRANET as a "global network, trading recipes and technologies from Maine to Nepal."[52] Ellis was trained as a physicist and was a former science adviser at the US National Science Foundation and at UNESCO. He began *TRANET*—a quarterly newsletter in which members of the network sent in ideas and experiences with appropriate technologies around the world and he then compiled them and sent them back out to the membership at large—after meeting Schumacher, first in the 1960s at a meeting on technology for the Third World, and later at the UN Conference on Human Settlements in 1976 in Vancouver, known as Habitat I.

In a Schumacher lecture he delivered in October 1998 in Salisbury, Connecticut, Ellis quoted a contemporary report from the *Washington Post* to capture the essence of Habitat I: "There was lots of hair, lots of blue jeans, and lots of protest. Yes, but there was a great deal more than that. The official delegates downtown with their army-chauffeured cars had their briefcases full of documents prepared for their foreign ministers, but the citizen representatives in blue jeans had the expertise. They made some sparks that illuminated a few important matters, the most important being that people aren't the problem, people are the solution."[53] Ellis liked the latter phrase so much that it became TRANET's motto.

Habitat I workshop participants discussed setting up an international mechanism for appropriate technology at length. Ellis recalled:

> I don't know how many meetings I went to—they were held every other week, it seems to me—in Vienna and Paris and London to talk about establishing such an international mechanism. But people in the workshops who had come from the Third World and from the grassroots and who were developing their own technologies rejected soundly the idea of a UN organization telling them how to develop what they needed at the local level. They feared that a centralized bureaucracy would override citizen participation, and they said: "All we want is to stay in touch with one another."[54]

George McRobie, who had helped Schumacher found the Intermediate Technology Development Group, suggested that Ellis be the facilitator because, in his role at UNESCO, he had been instrumental in bringing people together from around the world. Shortly thereafter, Ellis and his wife left Washington, DC, and moved back to Ellis's childhood home in rural Rangeley, Maine. There, they had "a woodlot for heat, a great big garden, and maple trees from which we could make maple syrup and maple sugar." They added a greenhouse and decided to live the "frugal life with our own appropriate technologies."[55]

Ellis resisted TRANET's classification as an NGO, however. He preferred the term *GRO*, for "grassroots organization." Reflecting on his time at the NGO Forum in Vienna, he said that NGOs associated with the United Nations "are essentially part of it. At the Conference on Science and Technology we found that the official NGOs waited for the governments to come to them and tell them what to support. There was very little original thinking in that particular group of NGOs." Rather, TRANET was about "people at the grass roots doing it on their own."[56] For example, when Ellis began searching for an appropriate waterwheel design for Lamjung, a village in Nepal he had visited, he was able to connect the Nepalese villagers with Ron MacLeod, an acquaintance from Rangeley with expertise in the historical waterwheel designs of New England who was able to propose "a once-famed, but forgotten, turbine built for landscapes not unlike those of Nepal."[57]

In 1982, Richard Harley of the *Christian Science Monitor* latched onto TRANET's futuristic appeal. "These days, linkups like the Rangeley-Nepal connection are far more than haphazard conversations. They're part of a conscious wave of global 'networking,' a revolution of social spidering that is weaving webs of planetary cooperation that were inconceivable even a decade ago," he wrote. And while there had been a long history of human cooperation through exchange, "whether by messenger or mail, smoke signal, or teleconference," Harley noted that "observers tracking the emerging global networks think a new page is being turned in the history of the social enterprise. The stage had been set for it in the 1970s. Planetary prophets were warning that the rising needs of burgeoning local populations could not be met by the overweight, slow-footed government bureaucracies. Communities in rich and poor countries alike searched for more local self-sufficiency. Many found a ready resource through networks of idea-sharing that linked up their regions. In the 1980s, networking has gone global."[58] In fact, although Harley and Ellis may not have been aware of them, VITA and the ITDG had set a precedent for exactly this type of global networking a decade and a half earlier.[59] Nevertheless, the increasing speed of globalization advanced by technology must have seemed exciting.

Of course, another form of networking was also taking off in the early 1980s.[60] In 1982, Ellis recalled, Apple gave him an Apple IIe computer "with a few of those five and half inch disks," and linked him with the Farallones Institute in California, Volunteers in Asia at Stanford, and the ecology department at University of California, Davis: "There were the four of us. Apple told us to learn how to network." With six-hundred-baud modems, "which meant it took you all day to write your name and get it transferred to any of these groups," they started networking via computer. They called themselves Eco-Net, for "ecology network," and by 1998 they had grown to ten thousand groups.[61]

In his retrospective address on TRANET, Ellis said that his task throughout his life "has been to search for the butterfly wings, as I call them, of science and technology." Drawing the metaphor from chaos theory, Ellis explained:

> The idea is that no matter how good a mathematical model you have for projecting the future of any complex system, the initial conditions of that system make it impossible to project very far into the future. The flap of a butterfly wing in Brazil may be responsible for a hurricane a year later in New England. Unless you can measure those initial conditions, you can't project the future. But most futurists look back at the trends of the past and say, "If we follow those trends in the future, then this is where we'll end." They don't take into account the effect of flapping butterfly wings on our world.[62]

Looking at the program of the NGO Forum in Vienna in 1979, one could say that the majority of panels were about the butterfly wings. Overwhelmingly focused on appropriate technology, the symposia and workshops conveyed a consistent message but refrained from engagement with the political apparatus embroiled in negotiations at the official UN conference up the street. At the same time, other NGOs working well within the systems of the US government and multilateral organizations' grants and contracts were at work developing their own versions of appropriate technology. One of those organizations was PATH.

∵

While the Ford Foundation gave PATH early seed funding for its work bolstering contraceptive supplies through linkages with the private sector, PATH's technology development projects were historically supported by multilateral organizations and government contracts. Interestingly, in an internal memo about a PATH grant proposal to his colleague, investment officer John Foster-Bey, on January 12, 1988, Bud Harkavy pointed out the foundation's historical lack of investment in technology development. "The problem of program relevance, however, arises," he wrote, "because we typically are not involved with technology development in our grant-making, hoping that appropriate technology will be developed and made available by others. My own sentiment is that we've gone overboard in our lack of interest in technology and I'd hope, if [program-related investment] funds were available and the business plan made sense to you, this proposal could be considered."[63] "Are you willing to put grant funds into this?" was Foster-Bey's one-line reply.[64] The proposal in question, for funds to put toward the development of new technologies for child survival, was one of the few PATH made to the Ford Foundation in the 1980s that was unsuccessful.

Child survival technologies were the order of the day in the early 1980s. Following the 1978 Alma-Ata Declaration, on primary health care, Drs. Julia Walsh and Kenneth Warren published what was intended to be an "interim strategy" on how to implement broad-based health systems. It focused on the four interventions known by the acronym GOBI—growth monitoring, oral rehydration, breastfeeding, and immunizations.[65] And while the Ford Foundation was not excited about funding new technologies in this area, PATH found enthusiastic partners in USAID and, to a somewhat lesser extent, the WHO. The USAID flagship HealthTech project, which is still ongoing, has been funded by USAID continuously since 1987. The WHO's support of PATH's oral rehydration and growth-monitoring projects makes intuitive sense—it was the body that came up with the Alma-Ata Declaration, after all. However, USAID's support for policies nominally in the service of primary health care may seem surprising, given the that on the surface they were at odds with the fiscal and political philosophies of President Reagan.

The story of the USAID-funded Technologies for Primary Health Care project, better known as PRITECH, is an illustrative encapsulation of these tensions. In spite of President Reagan's ideological opposition to the essential drugs program spearheaded at the WHO by Secretary-General Halfdan Mahler, USAID began many essential drugs projects in the early 1980s.[66] The essential drugs concept persisted because these programs used the policies and structures of privatization and cost-efficiency of the Carter administration to negotiate the friction between Reagan's bilateral policy and Mahler's multilateral one.[67] In a sense, the Reagan administration was able to have its cake and eat it too. PRITECH was nominally an essential drugs program, which allowed the United States to claim consonance with WHO policy. Yet it was conceptualized as a technology-transfer project, rather than a commodity-transfer one, which focused on a few select off-patent pharmaceutical interventions that posed no threat to the profits of American pharmaceutical manufacturers. The focus on discrete technologies meant that the projects were imminently quantifiable—numbers of vaccine and oral rehydration solution doses, numbers of growth-monitoring devices, and numbers of women given breastfeeding education. This fit with the administration's overriding concern with getting a return on investment for development aid, which some have derided as part of the larger neoliberal project of selective primary health care.[68] It also aided in making development projects legible between government agencies and contractors, enabling what is now called the "monitoring and evaluation" of government contracts.

The PRITECH project began on September 29, 1983, and was led by nonprofit international health organization Management Sciences for Health and contracted to a consortium which also included PATH, the Academy for Educational Development, and the Johns Hopkins School of Public Health. The original mandate of the project was "to improve primary health care (PHC) services and management capabilities through the transfer of selected, proven technologies, such as vaccinations and oral rehydration therapy (ORT), and through provision of PHC management and manpower training technical services to improve the design, implementation, and evaluation of PHC programs in the developing world."[69]

Less than two years later, however, PRITECH's ORT program was so successful and demand for ORT so great that the mandate was amended to focus solely on oral rehydration and the control of diarrheal diseases. Oral rehydration therapy is a simple glucose, salt, and water solution that is used to revive those suffering from extreme dehydration due to diarrheal diseases. Trialed during a severe cholera outbreak in Dhaka, Bangladesh, in 1967–1968, and used extensively for cholera sufferers in Bangladeshi refugee camps during the Liberation War, the formula was simple enough that families could prepare the solution at home.[70] It reduced deaths by a factor of 10 to 1, when compared to those who received only IV fluids. PRITECH's ORT program distributed the glucose-saline combination in dry packets that could be mixed into water at the correct ratio, eliminating room for error. As the memo announcing the program's scope shift pointed out, the contractors running PRITECH were "particularly skilled" in this area. This is true not only in the sense that the consortium had expertise in oral rehydration but also in the sense that they were skilled in disease control. Johns Hopkins School of Public Health was represented at the

PRITECH consortium by its then dean Donald A. Henderson, leader of the WHO's smallpox eradication campaign.

The project goals typified the selective primary health care approach—that is to say, even while engaging in a rhetoric of horizontal primary health care, the project supported a series of narrow vertical campaigns. In PRITECH's case, two of the four GOBI targets were covered under the original mandate—oral rehydration and immunization—and the focus was further narrowed to just oral rehydration with the amendment. The initial budget for the project, $19 million, grew quickly to $40 million after the first amendment in 1985, and again to $44.7 million in 1986. As it was considered a worldwide project (with operations in twenty-four countries), most of the funding came from the Bureau of Science and Technology in USAID, in Washington, with smaller portions contributed by the regional bureaus. The increase in budget was intended to allow the start-up of a second contract under the same umbrella (which became known as PRITECH II). It would focus on other aspects of health technologies, primarily immunizations, which had been dropped from PRITECH's original mission due to demand for oral rehydration. However, at the conclusion of the first six-year project in 1989, many of the oral rehydration activities of PRITECH were adopted by PRITECH II, which ran from 1987 to 1997, so that they could be sustained over a longer period. Soon, the immunization goals of PRITECH II were eclipsed by oral rehydration. The project therefore remained a single-issue, disease-specific campaign throughout its fourteen-year run.

One of the major challenges the PRITECH project faced was the supply of oral rehydration solution, an essential drug on the WHO's Model List. After "limited" commodity support was given by USAID and other donors—UNICEF contributed many packets in some countries, such as Morocco and Niger—project managers hoped the oral rehydration solution would be locally manufactured.[71] The project worked with the WHO and UNICEF toward this end in many countries; however, by 1990 project staff were looking to the private sector for assistance.[72] In countries such as the Gambia, Indonesia, and Kenya, PRITECH encouraged local governments to contract out to the private sector to maintain supply.[73] For the provision of health services, PRITECH also expanded beyond its public-sector engagement with the ministries of health in its host countries to the private sector, including private practitioners of medicine, missionary groups, traditional healers, and pharmacists.[74]

The PRITECH program typifies many of the contradictions involved in the circulation of pharmaceuticals in US health assistance in the late twentieth century. It entailed the mass transfer of oral rehydration solution, an essential drug and commodity, yet policy makers considered it a technology-transfer project. Although it was nominally in the service of primary health care, in practice it was an intensive vertical campaign, focused solely on the control of diarrheal diseases. Although oral rehydration solution was considered an appropriate technology, its local manufacture proved challenging and the project turned to the private sector. As with most USAID projects since the massive budget cuts and downsizing of the agency in the late 1960s and 1970s, PRITECH relied on private contractors to carry out the work. These inconsistencies paradoxically meant that the project could appeal to many different constituents and helped it survive not only the Reagan and Bush administrations but also the first term of the Clinton presidency.

PATH's HealthTech project, by contrast, aimed specifically to develop novel technological solutions that could be "scaled up" for transfer to the developing world. These devices included iterations of the VVM; Uniject, SafeTject, and SyringeLock single-dose injectable delivery systems that disabled the needle so it could not be reused; the PATHweigh and BirthWeigh antenatal and infant growth-monitoring scales; PATHstrips dipstick tests to measure a protein in an expectant mother's urine; and the PATHtimer and SteriTimer, small devices that when placed in a pot of boiling water or a pressure cooker, could let the user know when instruments had been sufficiently sterilized.[75] Many of these interventions did not go very far. Even with interested local partners and a low-cost manufacturer, successfully scaling up health technologies to the point that they can be consistently used by the people who need them has historically been the exception rather than the rule. Failure to scale, and the wasted research and development dollars it represents, is one of the most persistently critiqued aspects of technology-driven health and development projects.[76]

PATH's development model was speculative, akin to the research-and-development pipelines of pharmaceutical and biotechnology firms. Throughout the 1980s, however, it had limited funding to prototype new ideas. For many interventions PATH did design, it lacked the large-scale demand needed to get them off the ground. Its biggest successes came not from the research-and-development money that USAID provided through HealthTech, but through wide-reaching partnerships with the WHO and local governments. The VVM enjoys wide circulation because it has been mandated on all WHO and UNICEF prequalified vaccines since 1992. Uniject was successful in part because of funding contracts with the UK Department for International Development and the ability to fill it with any injectable substance. The historian Bill Muraskin has also written a celebratory account of the success of PATH's affordable hepatitis B vaccine in the early 1980s, which through a WHO task force and affordable manufacturing partners became a disease widely vaccinated against.[77] Their scope for experimentation—through increased funding for research and development and the expansion of field offices around the world that funding enabled—would increase drastically over the coming decades as the promises of computer technology began delivering. This was particularly true in Seattle, where, down the street from PATH's offices, Bill Gates's gamble on his own start-up, Microsoft, paid off in droves.

∵

PATH transformed notions of what was appropriate by leveraging partnerships with the private sector to make more advanced science and technology affordable and accessible for the developing world. However, the focus on these single point-of-use devices in American foreign aid can be a form of progress trap—reliance on small-scale, disease-specific interventions came at the expense of large-scale investment in sustainable infrastructure. The world focused on the continued, ceaseless treatment of children's diarrheal diseases with oral rehydration solution, for example, rather than the prevention of those diseases through the provision of pipes for clean water and sanitation facilities. That form of large-scale infrastructural investment hasn't been seen in foreign aid programs since budget cuts and appropriate technology programs pushed out the modernization theory driven forms of development of the 1950s and 1960s. As certain interventions succeeded (or didn't), the organizations governing global health moved on to solving the next major technical challenge.

PATH's conviction that the people of the developing world deserved access to the same advanced pharmaceuticals and medical care as the people of the West set a precedent for other NGOs that transformed ideas of primary health care. Perhaps most notable among them is Partners in Health (PIH), cofounded in 1987 by Dr. Paul Farmer, Ophelia Dahl, Thomas White, Todd McCormack, and Jim Yong Kim.[78] PIH grew out of a community-based health center in Cange, Haiti, called Zanmi Lasante. The organization originally focused on treating people with HIV/AIDS, at a time when most HIV efforts aimed at the developing world centered on low-cost prevention (through education and the distribution of condoms). PIH believed the poor deserved the same access to life-saving antiretroviral drugs as their richer counterparts in the Global North, as difficult and expensive as that was at the time. Paul Farmer and Jim Yong Kim (who would go on to become the twelfth president of the World Bank) similarly pioneered a treatment regimen for multiple-drug-resistant tuberculosis for patients in Haiti, Peru, and Russia.[79] They proved that cost-effective treatment was possible, so long as the infrastructure was in place to deliver antibiotics in an accessible, efficient way. As such, PIH builds hospitals and clinics and hires and trains local health care workers to run them in addition to their health care delivery work. The technology cannot operate in a vacuum.

It has always been tempting to search for technological solutions to the world's ills. Innovating out of a problem was often thought of as easier, at the outset, than tackling it at its roots. With fewer restrictions on how they could spend their time and money, NGOs and private companies had more leeway for experimenting with this type of innovation than previous generations of multilateral and bilateral foreign aid programs. However, critics were correct to point out the power dynamics involved in that trade-off. In the mid-1980s, as policy makers in the United States were moving away from programs aimed at increasing appropriate technology, the empowerment that small-scale tools promised began to appeal to other grassroots movements around the world. In apartheid South Africa and civil war–torn Zimbabwe, activists thought that power over human tools could be a source of liberation from oppressive regimes, which wielded large-scale technology and infrastructure in the service of colonialism and racist division. Maintaining that power, however, proved extremely challenging.

[CHAPTER SEVEN]

Bantu Technology

Comrade Joice Mujuru grinned widely as she cut the ribbon to officially open the appropriate technology demonstration village at the National Training Centre for Rural Women in Melfort, Zimbabwe, on June 21, 1989. It was a joyful affair. Mujuru, then minister of community, cooperative development, and women's affairs in the government of Robert Mugabe, gave a short speech thanking the government, nongovernmental organizations, and individuals concerned with the burden of women's work in development programs. The undersecretary in the Ministry of Political Affairs, Comrade Mao, then introduced the head of the center, Comrade Kamanga, with a fist of solidarity in the air. She enthusiastically took to the microphone to introduce herself, wearing a dress patterned with soaring birds, a metaphor, perhaps, for the rise of women in independent Zimbabwe.

The "appropriate technology demonstration village," as it was described on the cover of *Zimbabwe News: The Official Organ of the ZANU(PF)*, had been in the making since 1985.[1] Just as Gandhi and Nehru had advocated for the construction of model villages like Etawah in India during and after the independence movement, the Mugabe government had set about making plans to develop its own appropriate technology model village soon after it won the civil war for independence in 1980.[2] What set the village at Melfort apart, however, was its focus specifically on rural women. Mujuru played an instrumental role in the planning process. A veteran of the civil war, she had joined Mugabe's Zimbabwe African National Union–Patriotic Front (ZANU-PF) in 1974 at the age of nineteen, and soon became one of Mugabe's only women commanders. She was appointed a cabinet minister at independence in 1980—the youngest minister in Mugabe's new government. Mujuru remained within the top ranks of the ZANU-PF party for the next several decades.[3]

The village, on five acres along Mutare Road, included a showground and a four-room *musha* ("homestead" in the Bantu language Shona) built from locally sourced materials.[4] The unfired bricks and "improved" thatched-grass roof enclosed a cement floor, which protected against moisture and termites. The furniture, doors, and shutters were made by local women and were "all of low cost." There was a rat-proof granary, with a Blair pump to supply water. The "sanitation unit" consisted of a Blair toilet and solar heater made from repurposed truck inner tubes to give warm water for a shower. Next door was a kitchen garden, to be watered with wastewater in the dry season. It integrated chickens, rabbits, and bees and had plans to introduce milk goats. The grounds included a demonstration bush pump, as well as a variety of wood stoves, a twenty-loaf bread oven, and an outdoor classroom. Boarding facilities could house up to fifty-two trainees at one time.[5]

The ZANU-PF's influential Women's League, of which Mujuru was an outspoken member, ensured that women's development remained at the forefront of party politics throughout the 1980s. The National Training Centre for Rural Women, which the league helped establish in 1984, was radical in that it not only recognized that women were the primary users of many technologies disseminated in the name of international health and development but also sought to empower the rural women of Zimbabwe to use technology for their own personal benefit—to "lessen the heavy load women carry."[6] The village "aim[ed] to help communities (especially women) to improve their living conditions by using appropriate labor-saving technologies in self-reliant income generating projects," which would "call for equal and meaningful participation of community members in the decision making process."[7] The beneficial results would eventually percolate from the individual women through the village as a whole—a strategy that reversed the thinking of typical capitalist economics, which held that incentivizing business and free trade at the highest levels would "trickle down" to village-level schemes.

The focus on women and the rhetoric was timely—well within the United Nations' "Decade for Women" that had begun in 1975 after the first UN women's conference. That conference had acknowledged, in the section "Women and Children under Apartheid," that it was the black women and children under the racist regimes in South Africa, Namibia, and elsewhere who suffered most greatly, and it called on the international community to provide "moral and material assistance to all the bodies struggling to remove apartheid."[8] However, the explicit linking of women's work with labor-saving technology was quite novel. In a 1980 report on rural women for the International Labour Office in Geneva, the economist Martha Loutfi framed women as "the most silent participants in the economic life of developing countries," and noted that with technological changes, such as the green revolution and the creation of modern dairies, women's income and status declined relative to their male relatives.[9] She argued that the "new international order begins at home," a reference that pushed back on the male-dominated NIEO agenda to that point. While the NIEO negotiated for the transfer of high-tech, modern technologies in international forums, there was little recognition that it was predominantly women who did much of the work of development on the household and village levels—particularly in the health, agriculture, and water, sanitation, and hygiene sectors where appropriate technology was most influential.[10] This point was even further neglected in Western donor countries, where appropriate technology as defined by Schumacher was supposed to be small-scale, low-tech, and labor-intensive, so as to maintain rural employment and prevent urbanization. The village at Melfort, Zimbabwe, was significant, therefore, in two ways—its focus on women and its definition of appropriate technology as labor saving.

As nongovernmental organizations like the Program for Appropriate Technology in Health and the Intermediate Technology Development Group expanded their overseas field offices and began exporting their own versions of appropriate technology, more indigenous efforts like the village at Melfort were gaining ground in Southern Africa.[11] For these countries, appropriate technology was a means of self-help and community development that would, in theory, help reduce dependence on colonial powers, as in Zimbabwe, and oppressive regimes, as in apartheid South Africa. These localized efforts, while representing a fracturing of the appropriate technology movement that had developed within multilateral and bilateral foreign aid organizations, also served to give it a new meaning and purpose that sustained it during the later 1980s and 1990s, when major international donors had largely moved away from appropriate technology programs.

Self-sufficiency as a concept, like appropriate technology, is a double-edged sword. Empowering on the one hand, on the other, the concept of self-sufficiency has historically been used by Western governments and imperial powers as an excuse for curtailing contributions to, and shirking responsibility for, development programs. As an actor's category, however, the term deserves serious consideration. Both South Africa and Zimbabwe, most likely because of their respective independence struggles, were outliers in the broader history of appropriate technology and primary health care, which saw the popularity of the concept peak within donor agencies in the late 1970s. In part, this is due to the persistence of the state and the impenetrability of foreign aid in the region during the 1970s: apartheid South Africa, due to the oppressive, racist regime, was not a recipient of US foreign aid in the 1970s and 1980s; Zimbabwe was in the midst of a civil war for independence for much of the decade. The postcolonial Zimbabwean state did not easily cede to international NGOs. While appropriate technology faded from the international health discourse in the 1980s in much of the world, it was enjoying a prolonged period of attention in Southern Africa in the immediate post–civil war government of independent Zimbabwe and the antiapartheid movement in South Africa.

Defining appropriate technology on their own terms was for black Southern African leaders therefore a way of asserting their own vision of modernity and version of the future. David Engerman, among others, has framed development as the application of science to human and social problems, and the control of development resources, including the mediation of the availability of technologies, such as pharmaceuticals, as a way of fueling international contests for power.[12] In the sense that "science" is often popularly understood as politically neutral, this characterization aligns with James Ferguson's critique of development as an "anti-politics machine."[13] Black Southern Africans recognized that technology was inherently political. Whereas Western donors were concerned primarily with safeguarding capitalist conventions like the trade balance and patent rights, leaders in Southern Africa saw how technologies designed primarily by white men in North America and Europe for use primarily by or on the bodies of black women and children in Africa conveyed the very gendered, racialized, and class-based power relations they claimed to help overcome. The question of which technologies were used for health and development programs was therefore critical to independence.

The delicate realpolitik of defining policy in the midst of independence struggles meant that black leaders in apartheid South Africa and in the vanguard of post–civil war Zimbabwe had multiple, often conflicting, approaches to appropriate technology programs. Their goals, however, were similar—technology, as elaborated in the first CASTAFRICA meeting of 1974, should be a liberating force, freeing black communities from the ties of colonialism and dependency.[14]

The programs set up in the name of appropriate technology for health in South Africa and Zimbabwe—namely, the attempt at model villages in Zimbabwe, and the Transkei Appropriate Technology Unit (TATU) and the National Progressive Primary Health Care Network in South Africa—and their legacies helped redefine the meaning of appropriate technology in Southern Africa. Women were central to all these programs, although their experiences are easier to parse for some organizations than others. It is worth acknowledging up front that the sources on many of these programs are deeply asymmetrical. The apartheid South African government cracked down violently on dissenters, so written records were a liability. The Mugabe regime wasn't much more receptive to critiques of its government. I have therefore had to reconstruct the history of these programs through interviews and reading institutional archives—where they do exist—against the grain.

∵

Appropriate technology was imbued with a nationalistic pride in Zimbabwe during the civil war and in the early years of the Mugabe regime. It was a familiar concept in the country throughout the 1970s, particularly in the fields of sanitation and medicine. There was an ongoing conversation in elite circles about the utility of appropriate technology for poor Zimbabweans, even if there was still debate over what exactly appropriate technology was. Speaking on behalf of, rather than with, the poor was an unfortunate consequence of the very hierarchical and racialized segmentation of expertise in the up-to-then-apartheid Rhodesia. In the field of sanitation and hygiene, many well-meaning (mostly white) physicians and health practitioners in Zimbabwe were frequent contributors to local medical journals, promoting all manner of intermediate and appropriate health technologies. Most interpreted appropriate technology, like their Western counterparts, as simple and low-cost.

For example, writing in the *Central African Journal of Medicine* in 1977, Dr. Peter Morgan of the Blair Research Laboratory, in what was then Salisbury, Rhodesia (now Harare, Zimbabwe), outlined how simple-to-engineer, low-cost latrines, wash basins, and water pumps could be used at the village level to prevent bilharziasis (also known as schistosomiasis). "The WHO acknowledges that the promotion of health is linked with the promotion of social and economic endeavours in which even the most backward people must play their part," he stated. Framing bilharziasis as "largely socially determined," he argued that most community leaders understood the importance of hygiene—they just did not have the right tools. "For the simple villager, confrontation with more sophisticated technology, familiar to those who have grown up with it, represents a bewildering experience," he wrote. "In the quest for health improvement in the less developed nations, the need for simple technology is essential. By considering the villagers' point of view, by providing less sophisticated but effective hardware that is easy to install, understand and maintain, the chances of success in drawing community interest and participation are far greater."[15] The "Blair pump" and "Blair toilet" of the Melfort model village owe their provenance to Morgan's lab. The pump became what is now known as the Zimbabwe Bush Pump Type B, the plans for which have been exported all over the world. It is considered a quintessential appropriate technology.

Dr. Morgan was heartily celebrated by the science and technology studies scholars Annemarie Mol and Suzanne de Laet in their classic 2000 article, "The Zimbabwe Bush Pump: Mechanics of a Fluid Technology," not only for his community-based design but also for removing himself from the control of the design. Branding him a "feminist dream of an ideal man," Mol and de Laet described how "he puts a lot of effort into dissolving—believing that it is precisely this which creates pumps that yield water and health in their Zimbabwean sites . . . Serving the people, abandoning control, listening to *ngangas*, going out to watch and see what has happened to your pump: this is not the line taken by a sovereign master."[16] While perhaps not a "sovereign master," in positioning the locals as "backward," it is clear that Morgan did not see them as his equals, either. Other scholars, such as Clapperton Mavhunga, have recognized that decentralized technological design situates everyday users—who adapt the technology to suit their realities—as designers in their own right.[17] The technology's accessibility and adaptability were what made it appropriate.

Another Zimbabwean doctor and lecturer in community medicine in Salisbury, Dr. Raymond Thomas Mossop, presented appropriate technology as a "lucky break." "High technology is appropriate to the consultant physician or surgeon," he argued. "Their expertise is useful to our patients and ourselves when needed, but sometimes they are coerced by the patient, the general practitioner, or their own blind dedication to their particular brand of technology, to overuse it. The payoff in appreciated life in such cases is doubtful. Appropriate technology for most of us is to sit back occasionally and consider what we have seen, and perhaps to count."[18] Citing the examples of Edward Jenner and cowpox, Norman Gregg and German measles, and Denis Burkitt and dietary fiber, Mossop believed that more serious contemplation and patient history taking was necessary to get at root issues of illness. In Zimbabwe, he said that many physicians were so occupied with treating acute infections that they overlooked underlying issues like protein deficiency in children that made them more susceptible to infection.

Moderate and pragmatic, Mossop also ran for political office in the 1974 general election. It was a tumultuous time for the country. Since the early 1960s, black nationalists had been fighting the ongoing Bush War against the Rhodesian authorities. Rhodesia was the illegitimate successor state to the British colony Southern Rhodesia, which had been self-governing since 1923. Not wanting to grant majority (black) rule in exchange for independence from the United Kingdom, in 1965 white Rhodesians made a "Unilateral Declaration of Independence" under the leadership of Ian Smith, the country's first Rhodesia-born leader since colonization. The move was swiftly denounced as treason by the United Kingdom. The United Nations declared that the "racist minority régime" was illegal, forbade its member states from officially recognizing the country, and imposed trade sanctions.[19] As the white minority had blocked the path to political negotiations for majority rule, the Bush War intensified. Two major black nationalist groups fought the Rhodesian Security Forces and, sometimes, each other: the Zimbabwe African National Liberation Army was the armed wing of the predominantly Shona Zimbabwe African National Union (ZANU) and fought out of bases in neighboring Mozambique, the Zimbabwe People's Revolutionary Army was the armed wing of the mostly Ndebele Zimbabwe African People's Union (ZAPU), who fought mostly out of bases in Zambia. Both groups received military aid from the Soviet Union, as well as training assistance from China and North Korea, which gave the white government some tacit support from the United States as it raised the specter of communism.[20] Meanwhile, in 1969, white voters approved a new constitution and established a republic by referendum, which severed the last remaining ties to the Crown when it took effect in March 1970.

Representing the mining town of Gatooma, Mossop ran for the Rhodesian Party, a moderate white opposition party that favored including black Rhodesians in internal politics. The party's founder and leader, Allan Savory, an ecologist and environmentalist, had defected from Prime Minister Ian Smith's party Rhodesian Front in 1972 to become the first and only white member of the opposition, citing his disagreement with the way the Smith government had been handling the guerrilla fighters. "The government simply panicked and is now antagonizing the local population by corporal punishment which affects the innocent," he said, "This is a guerrilla commander's dream. It is a sure road to destruction."[21] He also caused an uproar before the election in June 1974, by claiming, "If I had been born a black Rhodesian, instead of a white Rhodesian, I would be your greatest terrorist."[22] Smith's Rhodesian Front won every white seat that year, including the one for Gatooma. Nevertheless, Dr. Mossop remained in Zimbabwe and went on to become the head of community medicine at the University of Zimbabwe and the president of the College of Primary Health Care Physicians.

By the late 1970s, the Bush War became untenable for white Rhodesians. Many emigrated to South Africa to avoid military conscription, and the trade sanctions and mounting diplomatic pressure forced the Smith government to negotiate a settlement. While Smith agreed in 1978 to transition to majority rule, under the initial proposal whites retained control over vital state functions, like the police, security forces, and judiciary. Moreover, ZANU and ZAPU did not participate in the 1979 election. They did not support the power-sharing arrangement laid out in the proposed constitution. Fighting continued until Margaret Thatcher convened peace negotiations at Lancaster House—an agreement was reached wherein the Unilateral Declaration of Independence ended and Southern Rhodesia reverted to a British colony for several months until internationally supervised general elections could be held. In February 1980, Robert Mugabe's ZANU party won 63 percent of the vote, and the country was granted independence—and renamed Zimbabwe—on April 18 of that year.[23]

After Zimbabwean independence, some elite practitioners were more critical of interpretations of appropriate technology as simple and low-cost. The concept of essential drugs, in particular, proved a sticking point for some. Dr. Robert S. Summers, for example, analyzed the WHO's essential drugs list in the context of the needs of Zimbabwe, where he was professor and head of the Department of Pharmacy at the University of Zimbabwe and a deputy dean in the faculty of medicine. Summers emphasized that, although the WHO list could form some basis for a Zimbabwean formulary, "The WHO list is not an exclusive list and does not take account of the need for more sophisticated drugs."[24] This argument was based, in large part, on the fact that where the proposed Zimbabwean formulary and the WHO list overlapped, the five categories of disease considered were tropical infectious diseases (amoebiasis, bilharziasis, giardiasis, leprosy, malaria), tuberculosis, epilepsy, psychological conditions, and hypertension, whereas the top five leading causes of morbidity and mortality in Zimbabwe in the 1970s were avitaminoses and malnutrition, cancer, cardiac disease, cerebrovascular disease, and cirrhosis. More "sophisticated" drugs than were available on the WHO list were therefore needed to address many of these maladies.[25] Dr. Summers left Zimbabwe in 1983 as the Mugabe government was taking shape. He went to South Africa, still at the height of its antiapartheid struggle, and took a post as professor and head of the Department of Pharmaceutics at the Medical University of South Africa. Now the University of Limpopo, the Medical University of South Africa was established under the apartheid system to train only black medical professionals, who were not permitted in all-white institutions. Summers taught his version of appropriate technology and essential drugs there until 2006.

Mugabe built on the mixed interpretations of appropriate technology in his own policies. In *Zimbabwe News*, his party's official paper, the president made it clear that science and technology—adapted to the local conditions in Zimbabwe—were the key to independence, not only politically but also economically. "Without the people's consciousness to be masters of their own destiny," one article declared, "Third World economies may not become self-reliant."[26] In 1987, CASTAFRICA II was held in Arusha, Tanzania. Zimbabwe had newly joined UNESCO as an independent nation, and its delegation reported that it had formed the National Council of Scientific Research and the new Department of Technology to "spearhead technological development, particularly in the manufacturing sector" as part of the overall economic development of the country.[27] The appropriate technology demonstration village was to be part of this local manufacturing effort.

As with VITA, the Mugabe government used model villages as a way of marketing appropriate technology. The village in Melfort, which was funded in part by the Danish bilateral foreign aid agency DANIDA, was the first of these enterprises. The idea was that people, and women especially, would go to the model village for training and take the ideas and plans back to their homes for implementation, which would contribute to overall rural development. This attention to women can be seen as a corrective to critiques of development and the ways of "counting" women's work by feminist scholars such as the economist Esther Boserup. In her study *Woman's Role in Economic Development* (1970), she argued that economic and social development efforts, in general, led to the disintegration of traditionally gendered divisions of labor—to the detriment of women's economic and productive power.[28] With reference to a study of the "Bantu areas" of South Africa, she points out that only 32 percent of total "income" is monetary, with the remaining subsistence income in kind earned through barter or products or services provided and consumed within the family—this latter being primarily women's work.[29] At the opening ceremony, Joice Mujuru echoed this argument. Quoting Ela Bhatt of the Self-Employed Women's Association of India, she said: "We welcome technology that improves our living conditions but we do not want technology that snatches away whatever little work we have. We are rural women, spending half our lives fetching water, fuel, and fodder. We want them at our door steps. We are artisans, help us to create better tools for faster production."[30]

Nongovernmental organizations and donors saw Zimbabwean independence and the early years of the Mugabe government as an opportunity for expanding their operations in Southern Africa. The Ford and Rockefeller Foundations both invested heavily in the newly independent country, as they saw Zimbabwe as an ideal venue not only for development projects but also for reaching other nations in Southern Africa that had fewer resources—particularly as it allowed them to avoid politically "embarrassing" transit through apartheid South Africa.[31] The Rockefeller Foundation gave many grants to the University of Zimbabwe, while the Ford Foundation looked into setting up a Zimbabwe field office to handle Southern African projects that had to that point been managed out of the Nairobi office.[32]

The Intermediate Technology Development Group, Schumacher's foundation, established a field office in Harare in 1984.[33] One of the group's major projects, funded by the Ford Foundation, began in 1988. Cleverly titled Tinker, Tiller, Technical Change, the project was intended to incorporate indigenous technical innovations into larger development programs and to help foster small-scale industries.[34] The project sought to correct the "erroneous assumption" in technical assistance work in developing countries "that technological improvements can be devised and introduced only by outside specialists because local people are inherently resistant to change."[35] Rather, the project positioned "rural producers" as innovators in their own right and reframed the issue as one of inadequate documentation and visibility for indigenous innovations. Appropriate technology, for the ITDG, not only was developed with local needs in mind but also came from *within* the community—conceptually and materially.

Marilyn Carr, a development economist who worked for the ITDG throughout the 1980s, understood rural producers to be, largely, women. However, she criticized the introduction of technologies for rural women—such as water pumps and grinding mills—as, at best, having no noticeable impact on women's quality of life and, at worst, having a negative impact.[36] A 1992 report from Zimbabwe's Ministry of Community and Cooperative Development, *Building Whole Communities*, supported these findings. Women lagged men, it noted, in education, family life (including domestic violence), representation in media, and work. In the fourth volume of the report, focused on the problems women face, a comic strip depicts how commonly "distorted images of women distort development assistance."[37] One panel, on self-employment, shows a woman being offered credit by a man (presumably a banker or development officer) to start an "income generating project," on the assumption that "women are mainly housewives, so the projects can be based on domestic skills. We can give them little loans." The woman in question rests her head in her hand in exasperation and responds to the stereotype by saying, "I enjoy sewing and cooking but they are not the best way of making money . . . what we want are skills in business and technology and adequate credit."[38]

Zimbabwe's shifting economic reality was also apparent by the 1992 report. The seventh volume, *Organising for the Future*, was focused on selling the idea of structural adjustment policies. Perhaps the quintessential "neoliberal" development policy, structural adjustment programs were conditions imposed by the International Monetary Fund beginning in the 1980s on many of the countries—particularly in the Global South—that it had loaned money to, imposing cost-cutting measures like user fees for health care access and education. Framing structural adjustment as a "belt-tightening process, i.e., making sacrifices and spending less in order to save more money to invest in production," the report does not shy away from the fact that the policies would likely have the most adverse short-term effects on women and low-income communities, through increased unemployment, inflation and relative price increases, and social-service cutbacks and increased fees for services.[39] Nevertheless, the report also tried to position women and the low-income as the biggest long-term beneficiaries of structural adjustment, and the project as one of "growth with equity."[40] The overall goal was to be able to invest in more advanced technology for economic growth and development. In a clear pivot away from its model village strategy centered on women, by 1992 the Mugabe government, which claimed it elected to participate in structural adjustment (rather than having it imposed by a multilateral organization like the International Monetary Fund), believed that this high-capital technology was what would ultimately be most appropriate for Zimbabwe's future.

Mugabe's legacy in the country is complicated. He was praised as a revolutionary leader who helped throw off racist white-minority rule, expanded social programs like health care and education, and elevated the status of women in the country. However, he also privileged his own Shona people over other ethnic groups in the country. Frequently portrayed as a progressive gone awry, becoming tyrannical as he clung to power, these characterizations overlook his early animus toward minority groups in Zimbabwe.[41] After he came to power, he continued to fight ZAPU viciously in Matabeleland with his North Korea–trained Fifth Brigade, massacring over twenty thousand Ndebele civilians in what became known as the Gukurahundi, an attempted genocide that loosely translates from Shona as "the first flash floods that cleans the riverbeds of all debris and chaff before the spring rains."[42] His progressive vision was exclusionary, and became more so as fiscal austerity measures set in.

Comments on appropriate technology—its promises and failures—eventually found their way into the popular culture, as with the Harvard-, Columbia-, and Yale-trained Zimbabwean neurosurgeon J. Nozipo Maraire's epistolary novel *Zenzele: A Letter for My Daughter*. Maraire's work situates appropriate technology as part of a long lineage of short-lived development policies, imposed from the outside in and the top down. She wrote: "To them, we are the Third World, the backward countries, the developing, underprivileged world. Their agenda for us changes with each new trend in their own thinking, so that our course is never consistent.

One year we must generate a middle class; the next year they want to dump us with their technology; the following year they recognize that appropriate technology is the answer; and so it goes on." This takes on particular significance in the context of her novel's larger goal, which is to convey the experience of being a woman in Africa growing up in a newly independent state. Part history and part memoir, Maraire's work recounts the tensions of living in a nation in transition, looking toward the United States both as a land of opportunity and as a state imposing its will on those less powerful. Published in 1996, Maraire's novel joined a mounting chorus of critique of the appropriate technology movement, which had by that time lost its momentum.

∴

The Republic of the Transkei was carved out of the rugged southeastern coast of South Africa, a contrived black homeland for Xhosa-speaking people. Established by the apartheid government's 1951 Bantu Authorities Act as an administrative boundary for the black population, the unofficial state was the first homeland to achieve internal self-governance in 1963. The government of South Africa gave it a form of nominal independence in 1976, and the Xhosa were stripped of their South African citizenship and reassigned as citizens of what came to be known as a "Bantustan"—a place designated for speakers of the Bantu family of languages under the apartheid policy of "separate" (but not equal) development.

Despite the colloquially derisive origins of the term *Bantustan*, in South African historiography there has been a turn toward taking the artificial homelands seriously as a unit of study. Rather than approaches of the 1970s and 1980s, which tended to present polarized views of black South Africans as either supporting or resisting the Bantustans and the apartheid regime, recent historiography considers the nuances inherent in Bantustans as sites of political restructuring and coalition formation around the resources that were available in quasi-independent states like the Transkei. In her study of community health in KwaZulu, for example, Elizabeth Hull argues that the Bantustans provided both autonomy from the apartheid state and increased funds, which, though still inadequate, nevertheless allowed the KwaZulu government to create a progressive model of community-based health care.[43]

The black "homeland" of KwaZulu had been carved out of the province of Natal. The model there was based on the community-oriented primary health care clinic originally established in Pholela, a town in Natal, in the early 1940s. Started by doctors Sidney and Emily Kark, the clinic sought to "innovate, but inexpensively."[44] The Karks offered a comprehensive form of social medicine: combining free preventive and curative care, with an emphasis not just on treating the sick but also on maintaining the health of the community.[45] Health assistants would visit homes in the town to collect data on nutrition and the social and economic context of the family's health. This way, infectious diseases or diseases of nutritional deficiency would be caught early. The Karks' vision was one of self-reliance—families and the community took responsibility for their health in communicating with the health assistants and coming to the clinic for regular preventive care. The South African Department of Public Health enthusiastically adopted the Pholela Health Centre as a model, and between 1945 and 1948, it established forty-four similar health centers around the country. This was to be the foundation of a new national health system, until the National Party came to power in 1948 and enacted systemic racial segregation, known as apartheid.

The Pholela Health Centre continued to operate under the new regime, and the town became a site for further experimentation. Abigail Neely has shown how health became work that was enacted by women in the community—just as appropriate technology programs targeted women for programs on agriculture, water, sanitation, and hygiene, so too did they target women with educational campaigns about micronutrients and nutrition.[46] While the national health system the Department of Public Health had envisioned was abandoned under the apartheid system, the Pholela Health Centre model—known as community-oriented primary care—nevertheless inspired the primary health care movement at the WHO several decades later, which culminated in the 1978 Declaration of Alma-Ata.[47]

In the meantime, the African National Congress (ANC)—a group established in 1912 to advocate for black South African voting rights that resisted and opposed the apartheid government—took responsibility for the health of black South Africans. In 1960, after unarmed civilians protesting the Pass Laws (which required black South Africans to carry identification at all times and limited their movement in white areas) were killed by police in what became known as the Sharpeville massacre, the ANC and a break-off group, the Pan Africanist Congress, were legally banned. The ANC nevertheless continue to operate in exile out of friendly countries—first Tanzania, then Zambia by the late 1960s. It established a medical service, which became the Health Department, in the 1970s to serve the needs of the increasing number of exiled black South Africans living in camps in Angola, Mozambique, and other countries across Southern Africa. They received funding not only from communist allies but also from the WHO and other Non-Aligned countries.[48] The ANC used the diplomatic links the Health

Department generated to critique the apartheid medical system and put itself forward as a viable political alternative.[49]

The Transkei in particular was a site of strong resistance to the apartheid regime. As Timothy Gibbs has shown, the homeland formed what he called a "social hinterland" in which the young (primarily) men trained in the Transkei's elite ex-mission schools formed not only the backbone of the Bantustan's bureaucracy but also the activist wing of the nationalist movement as led by commissar of the ANC's guerrillas, Chris Hani.[50] It was through this regional, Transkei-based network that Nelson Mandela was able to rise to national prominence.[51] Contrary to characterizations of Bantustan leaders as being corrupted, or in league with the apartheid state, therefore, Gibbs's work on what he calls "Mandela's kinsmen" shows that many senior bureaucrats were working from within the Transkei to lead the ANC and the black nationalist movement to subvert the apartheid regime from underground.[52]

One way the leaders of the Transkei did this was through an embrace of appropriate technology, with the aim of increasing self-sufficiency. Sources from within the Transkeian bureaucracy are scarce; however, the Transkeian government hired Harvard-trained anthropologist Cecil Cook as a consultant in the early 1980s, who published widely on appropriate technology in the Transkei. Cook came to the capital, Umtata, to form the Transkei Appropriate Technology Unit, or TATU, in 1982, after experience consulting for the World Bank. His ideas were thoroughly Schumacherian. In the newsletter of the South African New Economics Network several years later, he wrote that "unless the rapid collapse of rural communities is arrested and reversed, it is only a matter of time before the hungry, ill-housed, under educat[ed] and unemployed citizenry of South Africa—trapped half way between the rural and urban zones of civilization—will cause the metropoles to collapse into chaos of apocalyptic proportions."[53] Like Schumacher, Cook advocated for development approaches that were low cost, that would help maximize employment in rural areas, and which would discourage urbanization.

He thought this could best be achieved with simple, locally sourced technologies that would increase independence and autonomy from the centralized South African state, and he hired community organizers—often women—to help spread the message.[54] TATU's mandate included developing medicines from edible wild plants, building homes that leveraged architectural and design elements to improve sanitation, and using passive solar heating to improve agricultural yield and reduce costs per calorie of feeding the population. However, these projects proved controversial within the Transkei.

According to Cook, the Transkeian elite had no interest in a vision of the future based on Bantu technology. In an article, titled "Pie in the Transkei," Cook (who fashioned himself a "social anthropologist, appropriate technology technocrat, and solar builder") laid out what he thought was preventing TATU's success. While appropriate technology as espoused by TATU was one of "self-determination through economy" this vision, according to Cook "simply did not project the kind of future that the Transkeian elite desperately wanted to picture for themselves and their rural relatives," while First World design professionals were not keen on ultra-low-cost technologies, he noted, because their fees were based on a percentage of the total project cost.[55] Cook characterized the issue as one of marketing TATU and its approach, which the organization attempted to do by building dozens of demonstration buildings in rural communities. Unlike the model village in Melfort, Zimbabwe, these were not aimed specifically at women but rather at the mostly male decision makers. TATU hoped that the villages would generate interest and local participation in the design of low-cost and high-impact structures, although Cook frequently lamented that the Transkeian elites continued to prefer Western structures which were "technologically obsolete, environmentally inappropriate and economically unaffordable to the majority."[56] With local elite resistance to changing building design, TATU pivoted to focus on smaller-scale tools and projects that were locally sourced and labor-intensive—particularly for the women responsible for domestic work.

There are reasons to challenge Cook's account. The anthropologist Richard Rottenburg, who worked at the University of the Transkei in Umtata in the mid-1980s, recalled that Cook seemed either to not notice or to strategically bracket the delicate political connotation "appropriate" technology—low-cost, simple solutions—had in the context of the apartheid regime.[57] He tried to situate his work with TATU outside of the apartheid classification, in which low-tech solutions were appropriate for the homelands, while high-tech was reserved for white South Africa. However, even Bantustan leaders opposed Cook's interpretation and were intent on advocating for a more high-tech definition of what appropriate technology was. "For me, it felt a bit like cynicism to think about appropriate technology [as defined by Cook] in a Bantustan when in Cape Town and Johannesburg and Durban they were working on the high tech solutions for the survival of the apartheid economy," Rottenburg said. Lesley Steele, a community organizer who worked for Cook and TATU in that period, had similar recollections—that TATU did not always match what people in the Transkei thought was appropriate, often viewing Western ideas of appropriate technology as "not as good as."[58] As the head of TATU's community outreach branch, Steele said she would reach out to local leaders and chiefs to try to win them over to concepts like mud stoves and community gardens, but many would push back, asking, "Why are you trying to sell my people inferior technology?"[59] TATU lacked an adequate understanding of the multivalent and often conflicting desires of the Transkei's various constituencies—the governing elites, the more rural majority, and those engaged in Umtata's civil society.

At the same time, the government of South Africa was policing members of the Transkei's intellectual scene, particularly at the university. The University of the Transkei was an apartheid institution, created by the government just as it created the Bantustan itself. Yet as Rottenburg described, the faculty was quite international with a strong leftist inclination: "It was for instance men who ran away from conscription in the Republic of South Africa . . . and many [international faculty] came as leftists, to do something against apartheid. Quite a few came for the comparatively good salaries paid there."[60] Politically invested faculty members were routinely threatened, arrested, and sometimes deported by South African authorities, who closed the university down every few months. TATU, however, was independent of the university, and "ideologically, on a totally different track, namely 'let's do the best of this Bantustan'"—a notion "you couldn't even think of within the university without being put in the pro-apartheid corner."[61] TATU was, according to Rottenburg, never threatened by the Bantustan or the South African authorities. Cook came from the United States "with the perhaps somewhat naïve understanding of simply doing good," and he was careful not to run into trouble with government demands as his way of maneuvering around the apartheid regime's harassments.

This conformity could be read as complicity—Cook accepted compromises to keep himself out of trouble, while other foreign nationals working in Umtata used their positions of relative privilege to boost the antiapartheid resistance.[62] However, Rottenburg thought that Cook simply believed in the possibility to support marginalized and impoverished people by giving them the means to sustain their lives more independently. Rottenburg had to leave Umtata after one of his students was involved in blowing up the city's power station on June 25, 1985.[63] Several days later, the student was executed by the South African security forces, many of Rottenburg's colleagues were arrested, and he was placed under observation, as one of the faculty contacts of the ANC students' group in the Transkei.[64] A few months later, he was advised by the university rector to resign and leave the country or face deportation.[65]

Perhaps because Cook was so adept at staying out of the politicking between the Transkei and South African governments, TATU enjoyed rather impressive longevity. Renamed the Eastern Cape Appropriate Technology Unit (ECATU) after the fall of the apartheid regime, it lasted as a separate entity until 2014, when it became a part of the Eastern Cape Development Corporation, funded by the provincial government.[66] Cook would go on to teach at the University of Fort Hare, an historically black institution that trained many African elite, including Nelson Mandela, Desmond Tutu, Julius Nyerere, Robert Mugabe, Kenneth Kaunda, and Oliver Tambo. Many of the initiatives TATU promoted, such as the home-growing of the staple crop maize (rather than its commercial import—an effort to improve malnutrition by lowering the cost of food), were also successful in catching on for the long term, though, like Cook, were well-meaning but sometimes flawed.[67] The home-grown maize, in particular, may have come at an unexpected cost. The former Transkei has had one of the highest rates of esophageal cancer in the world for nearly fifty years, more than six times today's rate in South Africa as a whole.[68] The cancer epidemiologist Dr. Vikash Sewram, of the South African Medical Research Council and the African Cancer Institute, linked the incidence of esophageal cancer in the region to the high silica content of home-grown maize, which is introduced through the labor-intensive grinding process.[69]

∴

Khathatso Mokoetle's work on improving community health through advocacy and appropriate technology caught on more easily than Cecil Cook's. As the general manager of the National Progressive Primary Health Care Network, which she helped get off the ground in 1987, Mokoetle not only oversaw the national office in Johannesburg but also coordinated the activities of eight provincial offices across South Africa.[70] Trained as a nurse and epidemiologist, Mokoetle has spent her career working for women's rights, particularly around health care, reproduction, and sexual violence.[71]

While the Transkeian experience seemed emblematic of the failures of appropriate technology in South Africa (one interlocutor told me, "Mandela is from the Transkei—if it didn't work there, where will it?"), grassroots organizations like the National Progressive Primary Health Care Network were more successful at articulating a version of appropriate technology that appealed to the local population.[72] TATU was top-down and led by a white American man, the network was bottom-up and led by a black South African woman. Unlike previous appropriate technology networks aimed at international development, this one was an indigenous effort. The national office's staff were led by Mokoetle as general manager, Dr. Irwin Friedman as director, and many black women in key executive roles. That representation mattered. At a time when donors in the West had moved swiftly past comprehensive primary health care to "selective" primary health care, the network consciously invoked the Alma-Ata Declaration and health for all in its mission statement, which was written collaboratively by the three hundred delegates to its first national council meeting in September 1987.[73]

As in the Alma-Ata Declaration, the network's delegates saw appropriate technology as a key pillar that enabled primary health care, especially for community health workers. Focused on the structural and socioeconomic determinants of health and disease, the network's mission was explicitly concerned with health equity and bolstering vulnerable groups—women, children, the elderly, the disabled—to ensure accessible and affordable comprehensive health care.[74] Technology was thus not meant to supplant social medicine but to complement it. It was a tool in the toolkit, not the only one.

The network envisioned health care rooted firmly in the community—a return to the Karks' model in Pholela—with district health centers and integrated service provision from primary to tertiary levels of care within "a coherent health system."[75] "People in the Community should be involved in their own health care. They should be involved in both the planning and running of all health programs. This is the only way in which the needs of the people can be met," network members wrote in a promotional brochure.[76] They also worked closely with civil society groups, including trade unions and women's and youth organizations. In defining primary health care, the network members wrote:

> Primary Health Care (PHC) is not just what you do in health but also the way you do it. It must start with community participation as without that there cannot be true PHC. To get the community actively involved in health means that health workers need to listen to the local community they serve. PHC is also about equity and all communities in South Africa need to have equal access to basic health care. It was agreed that primary health care meant linking health care to larger issues like social justice, access to clean water and toilets, adequate food production and education. It meant breaking down the barriers between different sectors of development like agriculture, town planning, education and health and working together.[77]

The focus on equity and access was particularly relevant to black South Africans living in the townships and rural areas, where clinics were not easy to reach.

In 1988, the network became part of the Mass Democratic Movement in South Africa, which was an antiapartheid coalition formed in response to the apartheid government's crackdown on the United Democratic Front, the Congress of South African Trade Unions, and similar groups. In particular, the network's members opposed "the damage apartheid in health imposed on the people of South Africa," including the fragmentation of care along racial divisions, massive disparities in health and wealth, unequal access to health care services, a private sector "prized" while the public sector was "despised," and the promotion of tertiary care at the expense of primary health care for all.[78] Network leaders were harassed, questioned, and arrested by South African police for their involvement in the Mass Democratic Movement protests, but they described their involvement with the government from 1991 onward—when it was clear there would be a transitional government—as collaborative "preparation for governing."[79] They began several "seed" projects to initiate local, community-based primary health care systems, trained community health workers, and engaged in networking events to learn from shared experiences.[80] Network members hoped that this groundwork—based on a truly equitable, community-focused, participatory vision of primary health care and appropriate technology—would enable the postapartheid government to quickly enact a new health law and reform the segregated health system. However, this hope relied on the ability to convince the new postapartheid government that a community-based interpretation of appropriate technology was the answer.

In the early 1990s, as South Africa transitioned toward the end of the apartheid regime and looked forward to state building, the National Progressive Primary Health Care Network, under the leadership of Khathatso Mokoetle and Irwin Friedman, took on a vital role not only in policy development in anticipation of the new South African constitution but also in the provision of training and care, particularly for HIV/AIDS. It was also successful in winning funding throughout the 1990s, first from the W. K. Kellogg Foundation and the Kaiser Family Foundation, and eventually from bilateral donors like USAID. Its National AIDS Programme began in 1990 after the Maputo Conference on Health in South Africa, wherein the major health advocacy groups of South Africa committed to implementing the National AIDS Programme through the Network.[81]

The new postapartheid government did not necessarily make the network's work easier. As Friedman recalled, he went to congratulate Nelson Mandela's newly appointed minister of health, Nkosazana Dlamini-Zuma (then wife of Jacob Zuma, who would go on to become South Africa's fourth president), whom he had been friendly with for a few years. He remembers that the network was thrilled that it finally had a minister of health that members felt they could talk to and who supported health equity. Dlamini-Zuma did not share the same view of what would be appropriate for South Africa's new health system, however. She told Friedman pointedly that she was not interested in supporting HIV/AIDS programs or community health workers, which the network had been pushing as the basis for primary health care. Dlamini-Zuma had grown up in Natal and had herself relied on the Pholela Health Centre for care growing up.[82] She was supportive of the model, but for Dlamini-Zuma, doctors like the Karks (whom she consulted with in 1992 about setting up the postapartheid health system) should be at the helm. A higher level of care was most appropriate for the newly reinvigorated, liberated, idealistic nation. "What?!" Friedman recalled her saying during the meeting. "We will not have any second-rate community health workers in our system. We will have doctors on every corner, like Cuba."[83] Yet community health workers were already an integral part of the South African health system, and Friedman said he did not see how there wouldn't be space for them and doctors to both practice in South Africa. The demand was great.

To extend their reach, particularly to health workers and families in more rural areas, in 1997 the network launched a radio program called *Community Radio*. It was broadcast on actual community radio stations around the country and was accompanied by a monthly newsletter, also called *Community Radio*, which elaborated on different themes, whether the importance of visiting community health care workers for primary and preventive care or environmental health, mental health, and addiction.[84] *Community Radio* embodied the network's commitment to appropriate technologies—both the medium and the message were community based and focused on achieving maximum impact from relatively simple interventions. The same year that the program launched, South Africa passed a new health act that the network had been working tirelessly for since its inception. Although there was (and there remains) a gap between the legislative ideal and the everyday reality, in a human rights–based framework, South Africa became a country guaranteeing the right to basic health care for its citizens.

As the changes to the health system took effect, they were communicated to the South African people through *Community Radio*. "Phila," a member of a local health committee, and "Nolitha," a community health worker, were recurring fictional characters in the monthly newsletters and radio shows. The two women feature in educational spots designed to clarify how the new health system worked and what South Africans could do to keep healthy. In the October 1998 issue, Phila is pictured in a drawing, breastfeeding her baby. "Health begins with me," she says, as she goes on to explain: "In the past health services and resources were unfairly distributed. In some areas, especially rural areas, health services were limited or unavailable. With the changes in the health system, health services are moving closer to communities."[85] The rest of the issue details the new district health system and the levels of care available to all South Africans.

∵

Throughout the 1980s and 1990s, appropriate technology for health was redefined again and again in Southern Africa, particularly by and for women. Phila and Nolitha, communication technologies in their own right, were also, appropriately, women. Mujuru, Mokoetle, and Dlamini-Zuma had all, in their own ways, pushed for and interpreted competing visions of health care, modernity, and empowerment. And while appropriate technology programs met with mixed success in this period in South Africa and Zimbabwe, the legacies of these indigenous efforts far outlasted those of foreign nongovernmental organizations and bilateral donors. They laid the groundwork for understanding health as something to be intervened upon, primarily, through discrete technological interventions. In 2014, as TATU's successor organization, the Eastern Cape Appropriate Technology Unit, folded, the South African Medical Research Council partnered with PATH to create the Global Health Innovation Accelerator, a global health technology development incubator meant to further indigenous health solutions. Yet as with redefining appropriate technology according to African prerogatives, this remains an incomplete, underfunded shift in decision-making from the Global North to the Global South.

[CHAPTER EIGHT]

Scaling Up

> The architects, planners—and businessmen–are seized with dreams of order, and they have become fascinated with scale models and bird's-eye views. This is a vicarious way to deal with reality, and it is, unhappily, symptomatic of a design philosophy now dominant: buildings come first, for the goal is to remake the city to fit an abstract concept of what, logically, it should be. But whose logic? The logic of the projects is the logic of egocentric children, playing with pretty blocks and shouting "See what I made!"
>
> Jane Jacobs, "Downtown Is for People"[1]

Inscribed on the outside of a gray brick building, in large gray sans-serif all caps that matched the gray Seattle sky overhead, the words "Every person deserves the chance to live a healthy, productive life" greet visitors to the Bill and Melinda Gates Foundation's Discovery Center. It is a tall order, reminiscent of the goals of primary health care and the Alma-Ata Declaration of forty years prior, but tied to an essential tenet of traditional economics—productivity and work.[2] Inside the center—decidedly brighter and more colorful—are dozens of displays of inventive gadgets designed to solve particular health problems faced by the developing world. From water filters to vaccine coolers to baby scales, it was clear that the Gateses thought every person would get their chance at a healthy, productive life through technology.

Many of the devices were novel and interesting. The Janicki Omni-Processor, a multistage treatment system "takes human waste and creates clean water, electricity and building materials." Although the display indicated that the processor was being tested in Dakar, it was also clear that selling the idea of drinking water made from human feces would take a significant amount of work. A plaque next to the small scale model of the Omni-Processor asked, "Would you drink water made from poo?" and urged visitors: "Take our poll! Watch a video and see if Jimmy Fallon does it." Visitors could text "gates poo" to the number or scan a QR code. For the majority of visitors to the Discovery Center this is, undoubtedly, hilarious and edgy, like a dare. But it is easy to understand why people, especially those in and from the Global South, would be skeptical of drinking water made from poo—decades of colonial hygiene campaigns drove home the message that human feces were dirty, to be kept far away from sources of drinking water.[3] Using videos of celebrities like Jimmy Fallon (comedian and host of the Tonight Show) and Bill Gates himself drinking the water to prove it was safe is an engaging marketing approach.[4] It recalls 1892, when Max von Pettenkofer dramatically drank a flask of bouillon laced with cholera to prove to Robert Koch that the bacillus alone was not enough to cause disease. However, this instantiation relied on mobile technology for texting and QR scanning and a catchy, humorous phrase—it was widely reported as the "Gates Poo Machine" in the media.[5]

Many of the devices were also familiar interventions developed not by the Gates Foundation, but by PATH. The vaccine vial monitor, the conspicuous light-pink circle with the heat-sensitive square cut out of the center, was affixed to several empty vials on display. The Uniject system also featured prominently. That the Gates Foundation would claim these inventions as its own is not surprising—Gates and PATH have enjoyed a close symbiotic relationship since the mid-1990s, several years before the Bill and Melinda Gates Foundation was started.[6]

The focus on individual point-of-use technologies is in keeping with the principles that proponents of appropriate technology have espoused all along: tools bring progress, empowerment, and development. Tools and technology are supposed to be a leveling force, bringing equality to an unequal world. As Bill Gates himself wrote in the mid-1990s, "We are all created equal in the virtual world, and we can use this equality to help address some of the sociological problems that society has yet to solve in the physical world."[7] Access to tools is therefore framed as the overarching issue standing between poverty and plenty, illness and health. Yet this approach has come at the expense of investment in broader-based health systems that would prevent disease. Whole cities and villages would benefit from clean-water infrastructure, for example, compared to the relatively small number of families who could afford to buy a Janicki Omni-Processor, or maintain one that was donated. Investment at the community level has lost out to investment at the individual or household level. This is perhaps the extreme end of Schumacher's argument in favor of the small scale, one that he didn't foresee—taken too far, it erodes community rather than bolstering it. With most of the devices he invests in, it is clear that Gates *is* thinking small, but this is framed in the media—in podcasts like Silicon Valley's *Masters of Scale*, hosted by LinkedIn cofounder Reid Hoffman—as thinking big.[8]

Perhaps more than the technologies themselves, I was drawn to the back wall of the Discovery Center, where a timeline of the Gates Foundation's milestones lays out the institutional manifestations of this history quite clearly. Rotating wooden blocks engraved variously with dates and significant quotes flip around to show photos and captions highlighting major events. One, marked simply "1998" on the front, commemorated the $100 million grant to PATH that the Gates Foundation described as "the first of many." "Based in Seattle, PATH brings life-saving vaccines to millions of children worldwide not yet vaccinated against rotavirus, pneumonia, and hepatitis B," the back of the plaque read, over a photo of a woman in an outdated laboratory looking through a microscope. While it is somewhat curious that the description of PATH focuses solely on vaccine development, the more interesting part of the story is how PATH came to win that grant in the first place.

It is in many ways a Seattle story, borne of the city's activist, tech scene. In the early 1990s, PATH's vice president of administration, Suzanne Cluett, also served on the board of Planned Parenthood of Seattle. She had begun her work in international health as a member of the Peace Corps in Nepal in the mid-1960s, where she also met her husband.[9] After their term was up, both stayed in Nepal working for USAID for a few years before moving to Seattle, where her husband, Dr. Chris Cluett, began working as a scientist for Battelle in 1968. The first offices of PATH (then PIACT) were on the Battelle campus, and in the late 1970s Suzanne Cluett became one of PATH's first employees, working her way up to the vice presidency. The Cluetts lived near the Gateses—Bill Gates Sr. and his wife, the businesswoman Mary Maxwell Gates, who also served on the Planned Parenthood board.[10] Bill Gates Sr. called her "the neighbor lady" and, through Suzanne and Mary, eventually met PATH's cofounder and CEO Gordon Perkin.

Around the same time, Microsoft cofounder Bill Gates Jr. was looking for something meaningful to do with his fortune. He and his wife Melinda asked his father to help him find a cause to invest in. As Richard Mahoney, another PATH cofounder explained: "You know, Gordon is a very, very charming, very brilliant person. And so he has the ability to make friends with just about anybody." He and Bill Gates Sr. became great friends, meeting for lunches and dinners over many, many months. "And Bill Gates Sr. kept asking Gordon, Well, what do you think my son should do?" Gordon consulted with Rich Mahoney and they suggested children's vaccines as a particularly effective starting point in international health. Bill Gates Jr. liked the idea, and in 1995, his father hired Suzanne Cluett to begin setting up what would become the Bill and Melinda Gates Foundation.[11] Rich Mahoney cowrote the first grant application to Gates for a children's vaccine program. "We came up with a proposal for $7 million. We submitted it thinking this is the biggest proposal we've ever written in our lives,"

Mahoney recalled. "The answer came back, 'Well, this seems interesting, but don't you need more?'" So we doubled it. Again the answer came back, 'Yeah, well, just as interesting but couldn't you use more?' We increased the request to $45 million . . . which in those days was astronomical."[12]

Perkin continued to advise the foundation, and in 1998, PATH was awarded that $100 million grant. According to Mahoney, at the Gates Foundation's press conference to announce the grant, a press officer approached Gordon Perkin and said, "Oh, by the way, we're going to make the grant $100 million because we got to thinking about it and we couldn't imagine the richest man in the world announcing a grant that was, you know, earth shaking and then have it be only $45 million." Afterward, Bill and Melinda Gates invited around twenty leading experts in global health and vaccines to their home for dinner.[13] Among them were Gordon Perkin and Rich Mahoney, who remembered:

> Bill and Melinda kept asking, "Well, what do you think are the great challenges?" And so all of us in the room who had grown up in the day of being happy if we got a $20,000 grant or even at $100,000 grant, we were all very cautious. Finally, I raised my hand and I asked, "Bill, I think we may be missing the point here. Are you telling us that this $100 million grant is only the start? And that you are prepared to make available a lot more money?" He said, "Yeah." The atmosphere of the whole room totally changed. People started coming up with all kinds of ideas about what could be done.[14]

The conversation was inspiring. So inspiring that Melinda Gates recalled that she and Bill "turned to each other after dinner was over and said we're going to give so much more to vaccines. It was just natural for both of us."[15] The outcome, ultimately, was a $750 million investment in a new organization, Gavi, the Global Vaccine Alliance, launched in January 2000.

The nascent Gates Foundation was powerful, and as Anand Giridharadas has argued, networks were "the basis for much of this new power—networks that simultaneously push power out to the edges and suck it into the core."[16] The tools big technology companies made were democratic, in theory—Microsoft gave middle-class consumers the chance to own a computer, Airbnb and Uber let anyone start a "side hustle" by renting out a room or their car—but, in practice, also concentrated wealth and power among a few, core stakeholders. The Gates Foundation was formed through the personal networks of the Gates family, using money that had amassed through the near-monopolistic hold Microsoft held on the computing industry. The late 1990s were a difficult time for Microsoft and Bill

Gates Jr., who had been slapped with a major antitrust lawsuit in May 1998 by the US Department of Justice and the attorneys general of twenty states and the District of Columbia. His image suffered. As one analyst put it in a retrospective:

> The strategy during the three-day deposition was classic Microsoft. Obstruct. Paint the government as out-of-touch policy wonks who had no idea how tech and real markets worked. And above all, deny even the most basic of premises in the government's case. The plan from Gates' army of lawyers and PR handlers seemed to be to wield his image as a software wunderkind who dropped out of Harvard to bootstrap his company and went on to become the world's richest man. Team Gates planned to use that same domineering force of will to beat back government lawyers. By day 2, it became clear that strategy was failing spectacularly.[17]

For all the rhetoric around tools as an equalizing force for consumers, it is clear through the deposition that Gates envisioned this within his own somewhat narrow parameters. He had no desire to loosen control of his platform to enable other developers to compete, and he fought efforts to ensure compatibility so that he could maintain his company's competitive moat. After several years of unfavorable media coverage, the Gates Foundation helped to resuscitate Bill Gates Jr.'s reputation, and it relied on the provision of the same types of market solutions—commodities, with intellectual property protections, that were just within reach for many developing country consumers—that Microsoft had built its business on.

This had real-world implications for the vaccine initiative. As with the negotiations over appropriate technology at the UN Conference on Science and Technology for Development in Vienna back in 1979, intellectual property protections were a key sticking point for pharmaceutical manufacturers. In forming Gavi, Gates was able to create affordable vaccines within the capitalist economic system by guaranteeing manufacturers intellectual property rights and securing their profit margins. Manufacturing low-cost vaccines for poor countries was not lucrative work, and it disincentivized would-be vaccine producers. Rather than having poor countries buy small quantities directly from drug companies, Gavi stepped in to buy on behalf of many countries at sufficient volume that it would be profitable.[18] This was all happening in the context of a major controversy over intellectual property rights for HIV/AIDS drugs at the World Trade Organization (WTO). Antiretroviral drugs were unaffordable to the mostly poor, sub-Saharan African countries that needed them most, and under the WTO's Trade-Related Aspects of Intellectual Property Rights Agreement, they could not be manufactured or sold generically.[19] In November 2001, a settlement was reached in the form of the Doha Declaration, which allowed for compulsory licenses to manufacture drugs generically amid public health crises. Gates, through Gavi, found a way of securing vaccine affordability without sacrificing intellectual property rights.

Under Perkin's watch, the Gates Foundation adopted PATH's model of appropriate technology development wholesale. It identified a need, worked with local partners and manufacturers, prototyped solutions, and then attempted to both scale them up and keep them affordable. Like PATH, Gates created public-private partnerships, which are now ubiquitous in the global health arena. At the World Economic Forum in Davos, Switzerland, in January 2003, Bill Gates Jr. announced a $200 million dollar grant partnership between the Gates Foundation and the US National Institutes for Health to address what he called the "Grand Challenges in Global Health." The fourteen "challenges"—enumerated in October of that year after the foundation's scientific board publicly consulted with scientists and institutions around the world—focused overwhelmingly on the development of science and technology.[20] The technologies were not old, repurposed, or outdated, but rather completely new, designed specifically for the exigencies of the developing world. Although the initiative was criticized for focusing too heavily on technology and ignoring the social determinants of health, Gates had long before invested his faith in technology's ability to transcend social conditions.[21]

The money Gates could throw behind these initiatives meant that they could take significant risks with research and development—they were beholden to no funding or donor agency's oversight. The foundation and its original three trustees—Bill Gates Jr., Melinda Gates, and Warren Buffett—wielded an annual budget of over $5 billion, just shy of the WHO's $5.5 billion per year, to which it contributed around 12 percent of the total.[22] The scale of Gates's investment in health technology also meant that novel appropriate technology development went from being a model of global health to *the* model of global health, eclipsing many other efforts at strengthening health systems from the bottom up. Bill Gates Jr., advised by Gordon Perkin and others in the PATH sphere, became the major agenda setter in global health worldwide.

Gordon Perkin, PATH's CEO of thirty years, left the organization in 2000 to become director of global health for the Gates Foundation. Asked how he thought PATH's concept of appropriate technology had shifted over the years since its founding, Gordon Perkin replied: "Certainly hard technology, physical technology is less a central part of the program than it was when we started. We built the initial program around our technology, if you like. The vaccine vial monitor. Things you could touch and feel and use and apply, and that addressed significant problems. And I think they've gone a little bit beyond that now and are more into I guess what you might call soft technology."[23] He pointed to the Plain Talk program, to develop health communications materials for nonreaders, as an exception. In 2010, he received the Order of Canada, his native country's highest civilian honor, for his work in furthering global health. Perkin passed away in August 2020, but his legacy lives on.[24] For many years, PATH was the Gates Foundation's largest grantee. Today, it remains the third largest—after Gavi and the World Health Organization—having received billions of dollars from Gates's coffers.[25]

∴

For most visitors to the Gates Foundation's campus, the Discovery Center is the entirety of what they will see. The complex housing the foundation's offices is one building over, closed to the public. I was fortunate to have been granted an interview with Chris Elias, the second CEO of PATH and, since 2011, current president of the Global Development Division at the Gates Foundation. Waiting for my appointment in the towering lobby at dusk, I looked through a glass wall to an art installation like a giant illuminated fishing net, suspended over the courtyard. Crafted by Janet Echelman, of Boston, the piece, *Impatient Optimist*, was created "to evoke the mission and optimism" of the Gates Foundation. Described as "a physical manifestation of connectedness," the knotted fiber net's colored lights come on sequentially throughout the night as the sun rises in each of the foundation's offices in Europe, Africa, Asia, and the United States.[26]

Chris Elias knew PATH and Gordon Perkin well before he took over as CEO in 2000. An internist, he'd lived in Seattle on and off, completing his master's in public health at the University of Washington in 1990. He spent the 1990s working for the Population Council, an international NGO focused on family planning and reproductive health, and in that capacity he both collaborated with and competed against PATH on various bids and contracts. Between 1996 and 2000 he lived in Bangkok, where the Population Council and PATH's offices were located in the same building. He recalled that for most of the 1990s, PATH's money was coming from USAID. Then, in 1998, Gordon Perkin made a trip to Bangkok accompanied by Bill Gates Sr.—the latter's first trip for the Gates Foundation.[27] The Population Council also received some early Gates Foundation grants, so the trio spent a few days visiting the project sites. "I remember talking to Gordon about this new [source of funding]. You know, it's hard now to recall what a big impact the Gates Foundation had when it arrived on the global health scene twenty years ago. I mean more flexible money, the only donor with funding of the size comparable to USAID," Elias recalled.[28]

After Gordon Perkin was recruited to be the first executive director of global health at the Gates Foundation, the PATH board hired the same search firm to find his replacement at PATH. The firm selected Chris Elias. Coming off of his stint in Bangkok, he said: "I was kind of ready for the next thing and one of the interesting things is that, you know, it was kind of a leap of faith." In Bangkok, he had been managing a staff of twenty people and a budget of about $3 million. "And PATH, when I joined in 2000, had 250 staff and a budget of I think about $45 million. And then, twelve years later when I left, they had 1,300 staff and a budget of about $325 million. So it was a time of tremendous expansion."[29] The expansion of the early 2000s was mostly due to Gates money—around 2005, about 75 percent of PATH's budget came from the large donor, though in several smaller grants rather than one enormous one. Under Elias's leadership, in 2006 PATH began a concerted effort to diversify its funding. "So we restructured the DC office. We hired some people who knew how to compete for business development with USAID. And between 2006 and 2011 we went from about $10 million a year of USAID government funding to about $100 million a year."[30] The growth that the Gates Foundation enabled fueled PATH's growth with other donors—it gave the organization the critical mass to compete. Elias also focused on diversifying geographically. PATH's projects had, until his tenure, been heavily focused on Asia. They had only one African field office, in Nairobi. By the time Elias left PATH to take over Gordon Perkin's former role at Gates, there were an additional ten field offices in eight countries on the continent.

By 2000, Elias recalled, the "small is beautiful" discourse of appropriate technology was not common at PATH: "That was sort of a theoretical discourse about appropriate technology . . . when I was there it was more about how do we get products that could actually get used." There were several barriers they had to overcome with each technology—many familiar tenets of appropriate technology. One was cost. The technology had to be affordable above all. "In some ways it would have been better [to call it] the Program for Affordable Technology in Health because it was 'how do you bring the cutting edge of science and technology to the poorest, to solving the problems of the poorest people in the world?'" Elias remembered that at the time many people thought of appropriate technology as being "for things like wells . . . it was more about things you could afford that weren't the best but they were good enough and they were better than what people had. We never had that discussion at PATH. It was always about 'how do we get the latest science into a form that actually makes a difference, right?'"

Elias's vision was one of science for the people. He gave the example of the meningitis A vaccine, which in the 1990s, he said, was being made for college students and the military at $400 a dose. "How do you harness that science for people who live in the meningitis belt, where there's, you know, 250 million people at risk of meningitis every year? And yet, to do it, you have to be able to afford it." PATH had to design for "extreme affordability." Most products were also designed to be used with little training—the vaccine vial monitor shows at a glance whether a vaccine vial has been exposed to too much heat and Uniject automatically disables after use to prevent reuse. Elias did not see PATH as part of the Schumacherian appropriate technology tradition. Rather, when PIACT broadened its focus from contraceptives, Perkin had told me, he thought appropriate technology was a cool concept and PATH was a cool name, but he was not a Schumacher disciple.[31] He made appropriate technology his own.

Appropriate technology went through many permutations since Schumacher first assembled the concept from his experiences in Burma and India. Schumacher's original vision was one of technology as a means to uphold rural employment and fix people in subsistence agriculture. Technology was a means of self-reliance. VITA and the ITDG largely kept this ideal in mind as they attempted to put the theory into practice. Their strategy tried to connect rural villagers and development workers with engineers and blueprints from the United States and United Kingdom as a means to advise them on which technologies would be most appropriate. As appropriate technology was adopted into the US foreign aid program, however, the rhetoric around it shifted—appropriate technology became a form of "development," self-reliance through charity. Implying a form of change or progress, promises of appropriate technology made through international forums were interpreted very differently by different voting blocs. As the US government made it clear that there was neither budget nor political support for large-scale technology transfer, NGOs like PATH leveraged partnerships with the private sector to make more novel and sophisticated technologies affordable for the developing world. Appropriate technology then became something revolutionary—rendering certain drugs and devices accessible to many millions of people who could previously not afford them. It carried this potential as it was adopted by government and civil society groups in South Africa and Zimbabwe, which saw appropriate technology as a means of liberation from the colonial and apartheid regimes.

Although the model of small-scale, radically affordable innovation that the Gates Foundation and PATH maintain remains the same, the term *appropriate technology* has waned in the global health and development discourse. In some ways, this is a shame. Appropriate technology invited the questions, "What is appropriate?" and "Appropriate to whom?" We are no longer having those debates in the same ways that they occurred in the 1970s and 1980s, even though terms like *sustainable development* and *disruptive technology* are no less nebulous. Instead, the Seattle–Silicon Valley innovation model is sold as the obvious answer, not least because the wealth that innovation in the computing industry generated positioned those same innovators as some of the largest and most influential donors to international health and development programs. The evergreen hope that we can hack our way out of global health challenges is celebrated every so often in the media—particularly in the midst of outbreaks like COVID-19 or Ebola—in spite of over forty years of experience that shows that relying on the development and "scaling up" of disease-specific individual devices without the support of broad-based health care systems is, at best, a scattershot approach.[32]

∴

In 2014, ECATU, the successor to TATU, shut down. It was absorbed into a local government development corporation. That same year, PATH launched the Global Health Innovation Accelerator (GHIA) in partnership with the South African Medical Research Council (SAMRC). Located at the very back of its campus in Cape Town, the GHIA is the latest attempt at an "African" appropriate technology program. Intended to be a commercialization incubator for new health technologies developed by South Africans for use on the African continent, the GHIA is an entity that enables PATH to claim community buy-in and local participation, in order to bring in continued donor funding. Heavily celebrated on its website as a satellite lab to downtown Seattle headquarters, in reality, the GHIA is the part-time concern of three individuals who spend most of their day working for the SAMRC—in some ways the GHIA is a shell entity, in that there is no lab or real dedicated physical space in which to incubate new technologies. In fairness, the GHIA is still "scaling up" after a shaky start. PATH hired some part-time consultants to start up the GHIA in 2014. Although they'd identified a handful of biomedical technologies that they were attempting to help bring to market, funding was insufficient even with PATH as a partner. In late 2016, they received a three-year grant from the Gates Foundation, which helped them gain some traction.[33]

While the SAMRC funds and conducts health research with local funding from the government of South Africa, there was a gap in getting health technologies to market and commercializing them. In the partnership with PATH, PATH put the think tank together and funds product development, while the SAMRC funds and conducts the technologies' clinical validation. An example is the uterine balloon tamponade (UBT) to stop postpartum hemorrhage. Inserted into the uterus, the device quickly fills with water to provide pressure and stop bleeding. As the uterus regains muscle tone, the water is pushed back out into a connected bladder, giving maternal health practitioners a visual cue as to the mother's internal state.[34] Recently featured in the *New Yorker* as "reverse innovation," the UBT is an affordable, relatively easy way to save women's lives.[35] Western versions of the UBT were still relatively expensive, so through the GHIA, PATH and the SAMRC partnered with South Africa–based Sinapi Biomedical. Together the three organizations created the Ellavi UBT, an affordable version that is designed for use by any maternal health practitioner—inside or outside a hospital.

The market for the UBT should be large—Ellavi's website cites 14 million cases of postpartum hemorrhage around the world each year, resulting in 120,000 deaths and 11,000 hysterectomies.[36] It also has the potential to replace other small-scale health technologies deployed to prevent postpartum hemorrhage, like the Uniject filled with synthetic oxytocin and misoprostol, which have been used off-label to devastating effect.[37] After SAMRC-designed clinical trials in smaller health facilities where it was administered by nurses and midwives from 2016 to 2017, the Ellavi received regulatory approval in July 2020 for use in Ghana and Kenya.[38] Within six months of its launch, it was being used in more than thirty hospitals in South Africa across five provinces. Elizabeth Abu-Haydar, who managed the UBT project for PATH, was thoughtful in considering unintended uses for the technology. "When working with partners, or designing our own technologies, one of our key drivers is always ease of use and 'do no harm.' But the realities on the ground are always a bit different than in controlled trials or even operational research that does help inform our introduction strategy," she wrote.[39] SAMRC and PATH employees alike consider the Ellavi UBT a good example of an appropriate technology, because it works seamlessly with the current health system, does not require extra work or training, improves patient outcomes, makes the job easier, and is usable at the community health worker level.[40] The Ellavi, literally, saves lives.

Scaling up a technology to make it widely accessible is often more difficult than coming up with, prototyping, and trialing a brilliant solution. Dr. Tony Bunn, who was one of the consultants PATH hired in South Africa to start up the GHIA in 2014, worked for the SAMRC from 1994 until his retirement in 2013, although he continues to consult on a part-time basis.[41] He had always had an interest in commercializing tech. In 2004, within the SAMRC he began something he called the Innovation Centre. His goal was to reach out to the other SAMRC research units to make them aware of the potential to commercialize their research and to translate their discoveries into intellectual property and, hopefully, products. The South African government's Department of Science and Technology helped fund it as part of its broader innovation agenda.

The SAMRC, established in 1969 under the apartheid government, had functioned on what Bunn described as a "traditionally academic model." Staffed by PhDs and MDs, what counted were publications and other typical academic metrics like number of graduates and impact factor.[42] This changed in 2012, when the organization was restructured. The "change management" exercises involved reducing intramural activities and increasing extramural ones, with research units moved to "pockets of excellence" at South African universities.[43] Just before he retired, in 2013, Bunn transitioned the Innovation Centre into a new Strategic Health Innovation Programme (SHIP), which focused on offering research-and-development grants for novel and appropriate health technologies "with metrics such as proof of concept technologies, patents, and interactions with industry."[44] "This was a major shift in focus from the SAMRC status quo that I had been struggling to introduce since starting the Innovation Centre," Bunn explained. "On retiring, shortly after the launch of SHIP, Dr. Richard Gordon took over from me to drive SHIP which now accounts for roughly half of the grant funding budget of the SAMRC going to health technology R&D initiatives."[45] The focus increasingly has been on getting health technologies to market. Bunn was asked to stay on at the SAMRC for a year or two to help with the transition, and it was during that time, after several successful projects with PATH South Africa, that he pushed for an alliance between SHIP and PATH. "Both entities were involved in the same mission of developing appropriate and transformative technologies for low-middle-income countries and a partnership seemed a logical strengthening for both entities," Bunn recalled. According to Bunn, the funding landscape in global health has shifted since about 2014, with large donors like the Gates Foundation not wanting their money to stay in labs and headquarters in the United States and Europe. The GHIA solved that problem for PATH, enabling it to move some portion of its work to a site in the Global South.

However, PATH is still very top-heavy, an issue that has stymied the GHIA and its South African field offices. Bunn and others had wanted to spin off PATH South Africa as its own nonprofit entity, to have more flexibility with operations in the country—it could raise its own money, diversify donors, and potentially get its own dedicated grants from large donors like Gates without going through PATH's headquarters. PATH did not allow it. As some large projects came to an end, PATH's South African field office in Johannesburg had to reduce its long-term staff, from about sixty people down to ten or twelve when I visited in 2018—an unfortunately common result of the "projectification" of global health, with booms and busts following the two- to five-year timelines of discrete projects. Morale was low. However, Bunn and SAMRC employees I spoke to saw advantages to being in South Africa as opposed to the United States or United Kingdom, in that they were able to connect very easily with real local needs. Bunn used to "go sit with people," setting up meetings with the nursing sisters and just observing. One SAMRC employee recounted that there are doctors in South Africa who "just like to tinker," who come to the SAMRC to help them take their devices to market and scale them up. At the same time, the local market in South Africa would never be big enough to recoup costs on most technologies, so they constantly have their eye on international partnerships.

Most South Africans I spoke to were proud of and generally satisfied with their national health system; however, they had more ambivalent attitudes toward health technologies. It was clear that to most, it was the nurses and doctors that offered care, not the devices that they worked with, which were often suspect. One man in his late twenties told me he was skeptical of technologies, from X-rays to chemotherapy. "Aieesh. Where did it come from, all this cancer? When I grew up, all we had was TB and malaria. Now it's cancer, cancer, cancer, and diabetes, and high blood pressure," he said. "There must be someone creating these diseases. It's the only explanation. They say I have high blood pressure and I should take medicine EVERY DAY! I am TOO YOUNG to take medicine EVERY DAY! It's like having HIV. If I have to take medicine every day, what is the difference?"[46] For him, the everyday technology of a diuretic pill changed the experience of disease—rendering high blood pressure akin to HIV due to its treatment regimen. More broadly, he linked the proliferation of hospital technologies like X-ray machines to the increasing prevalence of cancer in the country. International partnerships to bring more health technologies did not impress him. Rather, the collaborations people most celebrated were the ones that centered on people, like the partnership South Africa has with Cuba to train South African physicians.

∵

Ernst Schumacher's overriding concern was that technology be of the appropriate scale. His foundation, the Intermediate Technology Development Group, or ITDG, changed its name to Practical Action in 2005. The organization's focus is still on small-scale technologies, but rather than calling them "appropriate" or "intermediate," it now advocates for what it calls "technology justice." Defined as "a world in which everyone has access to technologies that are essential to life, and technology innovation is centered on solving the great challenges the world faces today: ending poverty and providing a sustainable future for all," the goal has been framed as a way of meeting the United Nations' Sustainable Development Goals.[47] While it harks back to the NIEO and the goal of securing scientific and technological equity for newly independent nations, the irony is that those same initiatives were hampered at the Vienna conference in 1979, in part, by the discourse of appropriate technology, which was used to maintain the status quo. In insulating institutions like the United Nations and the WHO from NIEO demands throughout the 1970s, appropriate technology enabled neoliberal policy to take hold in foreign aid.[48] What was once a discourse of finding the right level of technology to achieve optimal development within the context of a given nation has become simply one of incremental common sense—"practical action."

Yet the goal of global health and development programs today is largely still one of developing to scale, if not the scale that Schumacher originally argued for. While mundane programming tasks have increasingly been farmed out to relatively low-wage settings, like the women in Kibera, Kenya, who program the artificial intelligence behind self-driving vehicles, Western media celebrate "lab hacking" in Zimbabwe, by which students equip their laboratories by building their own improvised equipment out of cardboard and water urns.[49] African women trained to program services for major Western profits are paid relatively well, by the standards of the Nairobi slum, but they are vastly underpaid relative to what it would cost to hire programmers in Silicon Valley. Improvised and self-built lab equipment may not be new and shiny and state of the art, but it's seen as better than nothing. In other words, African countries have long received technologies just good enough to be perceived as better than what they had before, but never *too* good—receiving shipments of outdated cast-off computers, expired or almost-expired drugs, and discarded medical supplies.[50] These initiatives are often framed in terms of reducing waste, recycling, and environmental stewardship. Reusing old things, in theory, reduces the need for new ones, even if overseas freight and land transfers leave their own environmental footprint.

As this story has shown, appropriate technology has made global health as we know it today. While some historians have pinned the transition from "international" health to "global" health on the rise of NGOs, tracing the history of appropriate technology demonstrates the plurality of actors governing international health from very early on, including not only the WHO but also bilateral agencies like USAID, large philanthropic organizations like the Ford Foundation, and private and voluntary organizations like the ITDG and VITA.[51] It also shows how a new generation of NGOs that began in the late 1970s, like PATH, reinterpreted and promoted visions of appropriate technology that differed greatly from those presented by state-based actors. Other scholars, like Allan Brandt, have attributed the rise of global health to the HIV epidemic.[52] However, it is no accident that HIV/AIDS control efforts in Africa have centered on the discrete distribution of small-scale technologies, whether antiretroviral drugs or condoms, rather than strengthening health systems as a whole. They were informed by an earlier logic—a model of global health, like that advanced by PATH, aimed at making innovative appropriate technologies that were "radically affordable" and easy to distribute; that were in fact designed to be used with no supporting health infrastructure. Similarly, models of disease activism as seen around HIV/AIDS echo earlier generations of activism around access to tools, both in constituencies in the United States and from within the African continent. Appropriate technology led many people to think of tools as the means to change the world—and they did and continue to save countless lives, which is worth heartily celebrating. But distributing commodities like oral rehydration salts, vaccines, and water filters indefinitely cannot make up for the underinvestment in the broad-based health care systems and infrastructure they were designed to bypass.

Thirty years on from Alma-Ata, in late May 2008, former director general of the WHO Halfdan Mahler addressed the Sixty-First World Health Assembly in Geneva. At eighty-five years old, in the midst of the unraveling of the global financial crisis, Mahler opened his address with a quote by author Milan Kundera: "The struggle against human oppression is the struggle between memory and forgetfulness."[53] In his speech, he went on to remind delegates of the "transcendental beauty and significance" of the WHO's constitutional definition of health as not merely the absence of disease but as a "state of complete physical, mental, and social well-being." Attaining this standard of health was "a fundamental right." Mahler chided delegates for falling well short of this standard of care, for letting the promises of "health for all by the year 2000" founder. "When people are mere pawns in an economic and profit growth game, that game is mostly lost for the underprivileged," he said. "It is, therefore, high time that we realize, in concept and in practice, that a knowledge of a strategy of social change is as potent a tool in promoting health as knowledge of medical technology."

For Mahler, holistic, horizontal frameworks, like primary health care, were still the answer. And though Mahler acknowledged that "it is much easier to be rational, audacious, and innovative when you are rich," he was careful to note that "the inspirational energies and evidence base" for primary health care came from the Global South. In concluding, Mahler called himself an "inveterate optimist" who believed that the struggle between memory and forgetfulness could be won in favor of the Alma-Ata vision. The phrase was of course reminiscent of Bill Gates's own self-description as an "impatient optimist." The difference is subtle but telling.

[EPILOGUE]

COVID-19

In late December 2019, as news reports about a new virus began to emerge out of Wuhan, China, the world watched and waited. Most thought it would be controlled quickly—at worst, they thought it might result in a few thousand cases, similar to the outbreak of severe acute respiratory syndrome (SARS), another coronavirus, in 2003. When it became clear that the outbreak was spreading in Europe and the United States, however, policy makers reached for technological solutions. First, ventilators; then, masks and test kits. Eventually, they'd affix their hope to a vaccine. It was a reflexive move, in keeping with decades of technological fixes for global health problems, but on an unprecedented scale. It was not just a pandemic response—it was a bid to save the global economy from ruin.

This techno-solutionism is at once pragmatic and shortsighted. Clearly, we needed small-scale technologies like masks to get the pandemic under control, and hospitals used myriad different machines and devices to keep patients alive. In many ways, it was awe inspiring to see how businesses rallied to support this life-saving effort by ramping up production, bolstering supply chains, and concentrating manufacturing on the most critical items. But as the technology commentator Evgeny Morozov argued early in the pandemic, the idea that every problem awaited the right technology had "transcended its origins in Silicon Valley and now shapes the thinking of our ruling elites. In its simplest form, it holds that because there is no alternative (or time or funding), the best we can do is to apply digital plasters to the damage. Solutionists deploy technology to avoid politics; they advocate 'post-ideological' measures that keep the wheels of global capitalism turning."[1] Containment measures that did not hinge on the distribution or sale of masks or tests, vaccines or ventilators—like social welfare to enable people to stay home and quarantine, grants to small businesses to enable them to keep afloat in spite of mandated closures, and longer-term investment in disadvantaged communities to rectify health disparities—required significant political will in a deeply polarized political climate.

In a clinical setting, technological solutions are short-term stop gaps that cannot address underlying structural determinants of health. The COVID-19 pandemic has once again exposed the deep-seated inequalities that have plagued both American society and the international political-economic system since their founding.[2] In the United States, Black and Hispanic people were dying at rates similar to white people a decade older—at upward of six times the white death rate in some age categories.[3] This was in part due to higher infection rates. People of color were more likely to work in jobs that did not allow remote work; less likely to have health insurance; more likely to live in multigenerational housing, which put older adults at increased risk because of contact with their younger, more exposed family members; and more likely to have preexisting conditions that exacerbated COVID-19's effects.[4] However, people of color were not hospitalized at correspondingly higher rates. Dr. Gbenga Ogedegbe at New York University's Langone Health Center has shown that once hospitalized, Black patients fared as well as or better than their white counterparts. "Existing structural determinants pervasive in Black and Hispanic communities may explain the disproportionately higher out-of-hospital deaths due to COVID-19 in these populations," Dr. Ogedegbe concluded.[5] To benefit from the technological solution, one needed to have access.

And that access has historically been jeopardized by neoliberal policy making. Without a comprehensive, broad-based health care system, hospitals and the technological salvation they provide are out of reach for many people in the United States. The situation was often worse in much poorer countries, where decades of structural adjustment programs gutted health care systems, and governments didn't have the money to invest billions of dollars in developing and purchasing vaccines. Morozov pointed out: "The critics of capitalism are right to see COVID-19 as vindication of their warnings. It has revealed the bankruptcy of neoliberal dogmas of privatization and deregulation—showing what happens when hospitals are run for profit and austerity slashes public services."[6] The pandemic quickly overwhelmed hospitals running on thin margins of personnel and medical supplies, even in some of the richest cities of the world.

Appropriate technology was not intended to be an austerity measure, but it was wielded as such just the same. In the COVID-19 pandemic, we grappled with the global health system that appropriate technology helped to make. Divorced from broad-based, community-oriented primary health care, the technologies alone could not overcome structural and social barriers to health. Meanwhile, the technological imperative and orientation toward more novel, high-tech gadgets has meant the neglect of supply chains for old, mundane supplies like surgical masks—resulting in a major shortage as the pandemic sharply drove up demand.[7] Moreover, focusing on access to technology detracted from classic public health measures that work to curb the spread of infectious disease, like contact tracing.

The pandemic has also drawn attention to the importance of health communications and the difficulty of wielding expertise in the face of overriding uncertainty and a constantly shifting scientific consensus. Changing guidance based on new data created mistrust in a public who couldn't understand why masks might be discouraged one day and mandated the next—unless the expert on TV was lying to them. As scientists mobilized to study and understand the SARS-COV-2 virus with record speed, the public was led on a series of panics over the availability of key technologies. In addition to the shortage of masks and personal protective equipment, early in the pandemic, the public watched as politicians jockeyed for ventilators for their constituents—with state governors often having to circumvent antagonistic federal leadership.[8]

The concern over ventilator shortages in the United States paled relative to the thrust of anxiety over ventilator availability in Africa. Breathless media coverage reported that South Sudan had four ventilators for 11 million people, the Central African Republic had three for its population of 5 million, and of the six machines in Liberia, one was the property of the US embassy.[9] Ten African nations had no ventilators at all. It seemed to spell impending doom. African commentators noted that ventilators would be ineffective anyway in health systems with insufficient supplies of medical oxygen, trained medical personnel, and durable power sources.[10] It was a critique of decades of underinvestment and neglect—like the refrigerators donated in the Democratic Republic of the Congo to maintain the cold chain for drugs and vaccines, the technology was useless without the basic underlying infrastructure to power and maintain it.

In the United States, teams of intrepid engineers and researchers from Villanova University, Georgia Tech, and University of California, Davis, hurried to design low-cost ventilators for use in low-income countries. "These machines include renovated ventilators from the 1950s, self-pumping bag masks and a device that can supply air to two patients at once," a *Forbes* reporter noted.[11] NASA's Jet Propulsion Laboratory designed its own modular ventilator prototype in thirty-seven days.[12] A group of engineers (and parents) in Maryland figured out how to reverse the flow of air on breast pumps and convert them into low-cost ventilators—and a major breast-pump manufacturer, Medela, took up the challenge.[13]

In South Africa, there has been a renewed push to increase self-reliance in medical technology. The country has had mixed success in making these initiatives sustainable.[14] In April 2020, South Africa launched the National

Ventilator Project to develop an indigenous ventilator model and committed to locally manufacturing ten thousand ventilator units for use on the continent. By June, they were up and running, and the target had been raised to twenty thousand units.[15] The South African health care system was nevertheless plagued with shortages of vital medical equipment. In response, the SAMRC's Technology Innovation Agency embedded a new Medical Device and Diagnostic Innovation Cluster (MeDDIC) under the GHIA in March 2021. South Africa imported 90 percent of its medical devices and diagnostics, and COVID-19 had highlighted the risks of relying on those supply chains.[16] MeDDIC aims to build indigenous capacity, working through public-private partnerships.

These efforts are all attempts at making new appropriate technologies, even if they don't use that term. Makeshift ventilators would perhaps have been an effective stopgap had COVID-19 treatment not rapidly evolved beyond their use in large numbers. But they do nothing to address the systemic inequalities that create the conditions in which poor people of color around the world more likely to die of infectious disease.

While the public worried about the availability of ventilators and test kits and sewed their own masks, Bill Gates was working behind the scenes on the type of solution that had originally animated his foundation: a vaccine. Dr. Anthony Fauci, now known around the globe as the director of the US National Institute of Allergy and Infectious Disease who had advised the Trump and Biden administrations on their response COVID-19, was another guest at Bill and Melinda Gates's Seattle home in the early 2000s.[17] They'd built a strong collaborative partnership with work on tuberculosis and trials for an HIV vaccine. As COVID-19 hit, the Gates Foundation was monitoring the situation from the foundation's office in Beijing. But *New York Times* reporters recounted from an interview with Gates: "If you ask foundation executives and Mr. Gates himself when it really got going, they would probably point to February 14, Valentine's Day. And on that day, top foundation staff arrive at Bill Gates' personal office just outside Seattle. And they get together for this working dinner. They bring in an expert from a nearby university who is showing them the modeling. And as Bill Gates told [us] . . . from this point on, we're on code red."[18]

Gates began calling high-ranking government officials and members of Congress personally, telling them to pay attention to the coming crisis. He gathered allies in the private sector. And he called his friend Anthony Fauci. As with the minuscule supply of ventilators in Africa, when it looked as though vaccine candidates were on the horizon, global health experts, including those at the Gates Foundation, worried about how poorer countries would possibly gain access. As Dr. Fauci recalled, Bill

Gates "approached it from the standpoint of wanting to fill him in, like OK, what's up? What's going on in Washington with the vaccine? What are you doing? But his main thing was he wanted to make sure, which has been a classic Bill Gates point, is that we've got to make sure that when we do the vaccines that it's the kind of vaccine that could be used in the developing world."[19] On May 15, 2020, the Trump administration formally announced Operation Warp Speed, a public-private partnership to accelerate the development and distribution of a COVID-19 vaccine. By investing in multiple promising vaccine candidates at the same time, the effort maximized its chances of success.

The initiative Gates pulled together came to be known as COVAX. In April 2020, the WHO (to which the Gates Foundation is the second-largest donor), launched the Access to COVID-19 Tools Accelerator, in partnership with the Coalition for Epidemic Preparedness (CEPI—which Bill Gates also helped found), the European Commission, and France.[20] It aimed to "accelerate the development, production, and equitable rollout of COVID-19 tests, treatments, and vaccines."[21] COVAX was the vaccine pillar of the effort, and it is coordinated by Gavi, the Global Vaccine Alliance (that Bill Gates established). The aim was to get 2 billion doses of COVID-19 vaccine to priority populations—including frontline health care workers and high-risk individuals—in low-income countries by the end of 2021. Through COVAX, donor governments pool funding to negotiate prepayment deals for vaccines and increase manufacturing capacity. In securing doses bought for their domestic markets, they are also able to donate surplus doses at cost to countries in need. The WHO wanted to play a leadership role in the initiative, but the Gates Foundation, CEPI, and Gavi blocked their effort. In response, the *New York Times* reported:

> When we talked to Mr. Gates about this, I mean, he said listen, of course, we're always talking to the WHO. And of course, they can and should and are playing a meaningful role in this COVAX initiative. But a lot of the work here to stop this epidemic, he told us, has to do with innovation—innovation in diagnostics, therapeutics, and especially in vaccines. And he made the point that this is just not really the WHO's realm. He was saying listen, this is us. This is what we're good at. This is what we have been paying attention to over the last 20 years.[22]

While COVAX is an innovative funding mechanism, Gates has faced criticism for his support of intellectual property protections for the vaccines developed. In April 2020, Oxford University was widely praised for promising to donate the rights to its vaccine candidate to any drug manufacturer, so that it could be provided at extremely low cost. The Gates Foundation urged the university to partner with a large pharmaceutical company, so that the vaccine could be brought to scale and, a few weeks later, Oxford signed an exclusive manufacturing contract with AstraZeneca that made no guarantee of low prices.[23] This has had significant consequences—in January 2021, it emerged that South Africa was paying AstraZeneca $5.25 per dose, more than double what European countries were paying for the same vaccine.[24] The Gates Foundation similarly declined to support an early initiative led by South Africa and India at the World Trade Organization, which would have allowed countries to temporarily ignore patent rights for COVID-19 vaccines for the duration of the pandemic.[25] It was clear that Bill Gates was relying on the model of public-private partnerships that he and PATH had originally negotiated, which held patent rights as necessary protections for innovation.

COVAX came to source most of its vaccine from the Serum Institute of India, the world's largest vaccine manufacturer and the Gates Foundation's long-standing partner for other vaccine campaigns. Before the vaccine had even proved effective through clinical trials, the Serum Institute had obtained a voluntary license from AstraZeneca to produce its own version for low- and middle-income countries, which came to be known as Covishield, on a mass scale.[26] While Oxford University provided the weakened adenovirus needed, AstraZeneca provided the technology transfer in the form of plans and the vaccine recipe. By the time governments started approving the vaccine for use in December 2020, the Serum Institute had hundreds of millions of doses ready to ship, and in April 2021 it was producing between 60 million and 70 million doses per month.[27] Initially, half of this production was promised to the Indian government, whose domestic vaccination campaign was the largest the world had yet seen, and which in turn donated vaccine doses to about seventy poorer nations. Covishield was particularly popular with developing nations, as it required less stringent refrigeration than the Pfizer and Moderna mRNA vaccines, and as it was produced generically, it was relatively inexpensive. In short, Covishield was a logical choice for COVAX. When the partnership was announced, Seth Berkeley of Gavi (which operates COVAX) said, "This is vaccine manufacturing for the Global South, by the Global South, helping to ensure no country is left behind when it comes to the race for a COVID-19 vaccine."[28]

However, by March 2021, as COVID-19 surged in India and people were dying en masse, the Indian government halted exports of Covishield. The Serum Institute could supply only the domestic market. This severely hampered the initiative, which could not deliver the promised doses to many countries, especially in Africa.[29] COVAX had bet big on the Serum Institute and did not yet have a diversified supply base, leading to what Dr. John Nkengasong of the Africa Centres for Disease Control and Prevention called a "vaccine famine."[30] As the pandemic eased in India, the government eventually loosened restrictions, and the Serum Institute began shipping doses to COVAX again in late November 2021. Demand proved unpredictable. Just a month later, with 200 million doses of Covishield gathering dust on its shelves, the Serum Institute stopped production. Concerned about waste, CEO Adar Poonawalla was quoted as saying, "I have even offered to give free donations to whoever wanted to take it."[31] His company had ramped up production so quickly that it was hit hard when demand and the sense of urgency around worldwide vaccination softened.

In June 2022, the WTO approved a five-year deal to suspend patent rights on COVID-19 vaccines, almost two years after the proposal was brought to the organization by India and South Africa.[32] The deal met with immediate criticism from all fronts. Pharmaceutical companies argued that waiving patent rights would undermine innovation, with industry lobby group PhRMA calling the waiver a "political stunt."[33] India, given the excess supply of COVID-19 vaccine on the market, argued that the deal took so long to negotiate that it would have little effect. Chris Elias, of the Gates Foundation, had argued throughout that intellectual property protections were never the major barrier to COVID-19 vaccine production in the developing world—technology transfer was.[34] The Serum Institute was able to produce AstraZeneca's vaccine because it received not just the blueprints but also precise instructions on the manufacturing process. Without those instructions, the patent waiver alone was unlikely to do much to develop indigenous production capacity. South Africa's government admitted as much in a statement, saying that "to scale up the production on the continent, further partnerships will be needed including access to know-how and technologies," although it praised the deal overall.[35]

∴

As the Gates Foundation takes on an increasingly large role in setting the global health agenda, it is evident that the era of US ascendancy within the Bretton Woods international system is drawing to a close. The retrenchment has been a long time coming—the US state began to hand over technical expertise and agenda-setting power to private contractors in the 1970s. But it accelerated under the Trump administration, marked perhaps most superficially by the withdrawal, once again, of US membership from the WHO and UNESCO (the Biden administration has since rejoined the WHO and announced intentions to rejoin UNESCO). As the United States shied away from collective European efforts to fund COVID-19 vaccine research to focus on its own Operation Warp Speed, its old Cold War foe, Russia, once again stepped up its influence in Africa by offering 300 million doses of its Sputnik V vaccine to the African Union.[36] But it is unlikely another nation-state will accede to global leadership in global health—the Gates Foundation is already there.

On May 4, 2021, Bill and Melinda Gates announced their divorce.[37] As they grappled with a very personal, human-scale problem, the world watched and waited to discover what it would mean for their foundation—and for global health. The Gateses are aware of the responsibility they carry. Early press releases reassured the public that the work of their foundation would not be affected by the couple's split. Thus far, that seems to be true, although there have been some changes in governance. Warren Buffett stepped down as a trustee of the foundation in June 2021, and in January 2022, four new board members were appointed—Gates Foundation CEO Mark Suzman; Strive Masiyiwa, the founder of Zimbabwean technology company Econet Group and the African Union's COVID vaccine envoy; Baroness Minouche Shafik, director of the London School of Economics and Political Science; and Thomas J. Tierney, cofounder and cochair of the Bridgespan Group, a philanthropic consulting firm.[38] Melinda, meanwhile, has agree to resign as cochair of the foundation in 2023 should the couple decide they can no longer work together. She has also diversified her philanthropic giving—in a pivot away from a joint commitment made with Bill in 2010 to give the bulk of her wealth to the Gates Foundation, in February 2022 Melinda released an individual Giving Pledge, noting that she intended to give her fortune away "as thoughtfully and impactfully as possible," citing her own philanthropy Pivotal Ventures, which focuses on issues affecting women and children in the United States.[39] With an endowment of over $50 billion, 1,600 staff members, and $5 billion in annual grants, the Gates Foundation's immense influence in global health seems impervious.[40]

For his part, it seems Bill Gates again has his eye on the future. A year after the divorce, he published a book entitled *How to Prevent the Next Pandemic*, an encapsulation of the lessons learned through COVID-19.[41] Perhaps unsurprisingly, a major theme of the book is the creation of "better tools," including low-cost vaccines, drugs, and diagnostics.[42] Gates owns the familiar criticism of technocentrism in the introduction, writing: "And yes, I am a technophile. Innovation is my hammer, and I try to use it on every nail I see. As a founder of a successful technology company, I am a great believer in the power of the private sector to drive innovation."[43] This book is about the historical antecedents that led to this strategy, in the hopes we can learn from the past. And Bill Gates may be thinking similarly. His first chapter is precisely about learning from our present, pandemic moment. "It's easy to say that people never learn from the past," he wrote. "But sometimes we do."[44] While tools and technology are a central focus for Gates, he also emphasizes the need to invest in health systems, the failures and global inequities of which were laid bare by COVID-19. Those failures have historically prevented many innovative tools from scaling up in a meaningful way and reaching the people who needed them most. With greater recognition of our past, I am optimistic that investments in global health can truly chart a path toward health for all.

Acknowledgments

I am touched beyond measure at the number of people who helped make this project possible. Foremost among them is my mentor, Jeremy Greene, whose careful guidance and support have indelibly shaped who I have become as a scholar. His quiet, unreserved confidence in me carried me through many periods of self-doubt and meant more than I can readily articulate. Randall Packard taught me how to closely study a subject that is often far away and the importance of seventh drafts. Angus Burgin's keen eye and sharp questions pushed me to think deeply about the connections I could draw across time and place. Abena Dove Osseo-Asare's astute commentary and wise counsel over the years improved this project in innumerable ways. Peter Redfield's immediate enthusiasm for the project gave me not only a fresh outlook but a renewed energy in the midst of all the writing. I owe a particular debt of gratitude to Kavita Sivaramakrishnan, Gregg Mitman, and Jeremy Greene (again), who read and commented on a draft of the entire manuscript, and to the two anonymous reviewers at the University of Chicago Press whose questions honed my thinking.

Many brilliant colleagues also read and commented on individual chapters, commented on conference papers, and discussed the book's overarching ideas with me. Deepest thanks to Robert Aronowitz, Michael Barnett, Sara Berry, Leyatt Betre, Anne-Emanuelle Birn, Amy Borovoy, Ted Brown, Sarah Cook Runcie, Julia Cummiskey, Vincent Duclos, Yulia Frumer, Jean-Paul Gaudilière, Clara Han, Julia Irwin, David Jones, the late Pier Larson, Kate Law, Vincenza Mazzeo, Ramah McKay, Marissa Mika, Graham Mooney, Kirsten Moore-Sheeley, Ayah Nuriddin, Shobita Parthasarathy, Anne Pollock, Jeff Reznick, Daniel Rodgers, Richard Rottenburg, Sandra Scanlon, Brad Simpson, Liz Thornberry, Heidi Tworek, Dora Vargha, and Mari Webel.

At the University of Chicago Press, I'm very grateful to Karen Darling for seeing the promise in this project and shepherding it through to publication. I am indebted to Fabiola Enríquez Flores and Katherine Faydash for their help in preparing the manuscript, and to Anne Strother and the rest of the University of Chicago Press team for bringing this book to life.

I had the great fortune to be a part of two very stimulating intellectual communities while I was writing this book. At Princeton University, my enduring thanks go to João Biehl, Arbel Griner, Sebastián Ramirez, Andrea Graham, Jessica Metcalf, Katja Guenther, Angela Creager, Pallavi Podapati, Samin Rashidbeigi, Julia Marino, Jay Stone, Mikey McGovern, and the entire History of Science Monday workshop. The Global Health Program staff have also been incredibly supportive of my work—thanks to Gilbert Collins, Justine Conoline, Sara Goldman, and Debra Pino Betancourt.

The Institute for the History of Medicine at Johns Hopkins University is a wondrous place to develop and grow as a scholar. The close community fostered by Randall Packard during his time as chair, and further encouraged by Jeremy Greene as he took over the post, has shaped my positive outlook on the academic enterprise more than anything else. Mary Fissell patiently and expertly taught me how to think about medicine as a historian. Daniel Todes taught me the importance of words, meaning, and metaphors. Gianna Pomata encouraged me to bring my audience into the historical moment. Nathaniel Comfort helped me to hone my craft as an interviewer. Graham Mooney has been steadfast, always, in his support and congeniality. I am indebted to Christine Ruggere for many things, but especially her skilled research assistance in tracking down records on some elusive old colonial doctors. Eliza Hill quite literally did a lot of heavy lifting on my behalf—managing the hundreds of books I ordered over the years (and saving me thousands of dollars in library fines). I owe her several more doughnuts. Coraleeze Thompson and her magic phone tree ensured that I was able to navigate the quagmire of the Hopkins bureaucracy with very few snags. Marian Robbins not only helped everything to run but also has been a friend, informant, and confidante. Danielle Stout was a cherished ally on the Homewood campus. I'm incredibly lucky to have Ayah Nuriddin as an academic BFF—she is an ebullient force and a joy to work alongside. She, and our many nacho dates, saw me through the many ups and downs. Julia Cummiskey has been an inspiration, a mentor, and a friend. The friendship of Kirsten Moore-Sheeley, Jessica Levy, Morgan Shahan, Jonathan Phillips, Joanna Behrman, Seth LeJacq, Eli Anders, Ada Link, Justin Rivest, Penelope Hardy, Emily Margolis, Emilie Raymer, Anna Weerasinghe, Emily Clark, Kristin Brig-Ortiz, and Alex Parry made Hopkins home.

I presented early versions of many parts of this manuscript at Hopkins's own History of Medicine, Science, and Technology colloquium, the Critical Global Health Seminar, the Twentieth Century Seminar, and the African Seminar. (Slightly) more polished versions were presented at the annual meetings of the American Association for the History of Medicine, the History of Science Society, the Society for the History of Technology, the Society for Historians of American Foreign Relations, the African Studies Association, the Society for the Social Studies of Science, and the American Historical Association, as well as at the STS in/and Africa pre-4S Workshop, the Speculative Futures workshop at the University of Pennsylvania, the "Considering the Counterculture: A History in Ideas" conference at Princeton, and the Princeton History of Science workshop. I am very grateful for all the feedback I received in these various forums. Jason Chernesky, Ezelle Sanford III, Tess Lanzarotta, Jenna Healey, Lisa Haushofer, Rosanna Dent, Robin Scheffler, Wangui Muigai, Susan Lamb, Mari Webel, Anne Pollock, Dora Vargha, Cal Biruk, and many other friends shared advice and encouragement, and made conferences much more fun.

I'm enormously thankful for the financial support I've received for this project. Grants from the National Science Foundation enabled me to travel to Seattle, London, and South Africa. I am grateful to Wenda Bauchspies for her talented management of this nervous first-time co-PI. The Michael E. DeBakey Fellowship in the History of Medicine, awarded by the National Library of Medicine and National Institutes of Health, provided not only access to Dr. DeBakey's incredible personal paper collection and the library's wider holdings but also the time I needed to process and really think through the material. Jeffrey Reznick went out of his way to make my time at the NLM extraordinarily productive, Stephen Greenberg ensured that I was well acquainted with the collections, Beth Mullen offered her editorial guidance for *Circulating Now*, and Christie Moffatt saw to it that I could do it all while wearing and nursing a newborn baby. The Samuel Flagg Bemis Travel Grant from the Society for Historians of American Foreign Relations afforded me the ability to travel to E. F. Schumacher's archives and personal papers at the Schumacher Center for New Economics in Great Barrington, MA. I would like to thank Amelia Holmes, the archivist at the Schumacher Center, and Susan Witt, the executive director, for their helpfulness, warmth, and camaraderie. A grant-in-aid from the Rockefeller Archive Center provided critical access to files from the Ford Foundation on PATH's early days, as well as several grants to the ITDG. At the RAC, I am indebted to Patricia Rosenfield, whose intimate knowledge of my research topic meant she was able to connect me with key players in the field, Tom Rosenbaum, whose knowledge of the RAC's collections is incomparable and whose company over lunch was always a delight, and Renee Pappous, whose help with all things made my multiple visits run smoothly. A travel grant from the Ford Presidential Foundation allowed me to visit the Ford Presidential Library in Ann Arbor and access a wealth of files relating to science and technology policy during his administration and beyond. I am grateful to Mark Fischer, Tim Holtz, and Jim Neel for their assistance throughout my time there.

This project would not be what it is without the insight of many of the people who have spent their careers developing and managing appropriate technology programs. At PATH, I am grateful above all to the cofounders, the late Dr. Gordon Perkin and Dr. Richard Mahoney, for their time and enthusiasm for the project. Kate Davidson ensured that my trip to Seattle was productive. Many PATH staff very kindly shared their insider's knowledge of technological development and global health priority setting. Elizabeth Abu-Haydar and Beth Balderston shared project-specific information that lent a concrete relevance to all I had learned. At the Gates Foundation, Dr. Chris Elias very generously and candidly shared his experiences working in the field, at PATH, and as head of global health at Gates. Everyone I interviewed in Seattle said that if there was one person I should speak to in order to truly understand global health as it radiated from the city, it was Chris. They were right. At the South African Medical Research Council, Dr. Richard Gordon and Dr. Shelley Mulder led me through their work with PATH and the Global Health Innovation Accelerator, as well as the SAMRC's own efforts at health technology development. Dr. Tony Bunn, retired from the SAMRC, was very generous with his time and insight. Socrates Litsios welcomed me to his home in Switzerland, where we discussed all manner of appropriate technology initiatives at the WHO. Alex Shakow, an alumnus of USAID, not only offered abundant insights of his own but also connected me with Tom Fox and Ann Van Dusen, who were equally helpful with their recollections of USAID "back in the day." Lesley Steele, Irwin Friedman, and Khathatso Mokoetle kindly answered all manner of questions about their work with TATU and the National Progressive Primary Health Care Network in South Africa. Rae Galloway shared her experiences working as a consultant for many global health organizations in the area of nutrition and her perspective on technological development. I owe much gratitude, too, to the men and women in Southern Africa who offered their insights and opinions on how appropriate technology programs had affected their health and that of their countries. These include Peter Rampora, Collins, Lufuno, Mandla, Thomas, and Sinethemba (who did not want their last names used), as well as many others who wished to remain anonymous.

I have benefited greatly from the help of many archivists beyond those already mentioned. At the Weill-Cornell Medical Center Archives, Lisa Mix very knowledgeably guided me through the Walsh McDermott papers and offered entertaining details about Father Hesburgh. At the Alan Mason Chesney Medical Archives at Johns Hopkins, Andy Harris and Phoebe Evans-Letocha patiently screened unprocessed materials on my behalf. At the World Health Organization Archives, Reynald Erard and Tomas Allen offered incomparable insight into how the collections functioned and how best to search for what I needed. Both were outstanding hosts during my time in Geneva. At the United Nations Archives, Amanda Leinberger provided swift and able assistance for my disparate requests. At the World Bank Archives, Tonya Ceesay guided my requests through multiple screening processes with equanimity. At the Wesleyan University Special Collections and Archives, Jennifer Hadley acquainted me with Douglas Bennet's personal papers and provided some helpful context about his time as president of the university. At the American Association for the Advancement of Science Archives, Norma Rosado Blake was incredibly helpful both in locating files and in scanning them for me. At the University of Witswatersrand's Historical Papers, Gabriele Mohale skillfully and efficiently directed me to relevant files. David Sepeke Sekgwele of the Wits' Adler Museum of Medicine generously showed me some of the institution's treasures. And thanks go as well to the friendly gatekeepers at Wits' Commerce Library, who let me peruse their collection of Southern African development economics journals to my heart's content (even without a student ID card). At the British Library, I was amazed at the unforgiving efficiency and relentless cheeriness of the staff.

Family and friends near and far have made this project possible. Jessica Morefield, Mike Casey, and my little niece Miriam provided a wonderful home away from home for the time I was in London. Bumping into Jenn Fraser made my visit to the WHO and Geneva much more fun. Visits to New York City archives would not have been the same without Dafna Rebibo and Cook Alciati's warm company (and guest bedroom) or without the cheery late-night visits to Julia Emanuel and Eric Feinstein (and baby Noa!). Craig Parker and Becky Walker (and Leia, Ben, and Theo) provided a friendly welcome to Cape Town and very sage advice on all things related to living and working in Johannesburg with small children. Dafna, Julia, Craig, and I met as master's students in the University of Edinburgh's global health program over ten years ago, where I suppose much of this project began. Major thanks are due to Charles and Linda Morefield, for the childcare burden they took on to enable me to be all the places I needed to be. My mom, Coby Schneider, still reminds me to look after myself. My dad, Richard Schneider, never wavered in his thinking that a PhD in the history of medicine was a good idea. My stepmom, Lan Wong, asked all the right questions and baked the best snacks. My brothers Erik and Nils, my sister-in-law Lis, and niece Alice reliably bring me back to reality and make summers in northern Ontario well worth the drive. My extended family has been an enthusiastic cheerleading squad.

I could not have completed research and writing for this book without the women who have competently and lovingly cared for my children while I was working, both at home and overseas. I owe an eternal debt of gratitude, above all, to my son and daughter's teachers over the years: Janet Hammond, Cynthia Collins, Margaret Moran, Angela Venier, Carey Helmick, Ruth Tapp, Nancy Butler, Annemarie Schoen, Lauren Thomas, Liz Wilson, Denise Coffin, Zarya Navarro, and Tricia Maher-Miller.

Finally, I am grateful to my son Leif (ten), and daughter Vera (six), for the light they bring to my life.

The views expressed in this book are mine alone; they do not represent those of my employer.

Notes

INTRODUCTION

1. See, for example, Paul Farmer, *Infections and Inequalities: The Modern Plagues* (Berkeley: University of California Press, 1999), 21.

2. See, for example, Tom Vanderbilt, "'Reverse Innovation' Could Save Lives: Why Aren't We Using It?" *New Yorker*, February 4, 2019. On the history of sustainable development and its entanglement with appropriate technology, see Stephen Macekura, *Of Limits and Growth: The Rise of Global Sustainable Development in the Twentieth Century* (Cambridge: Cambridge University Press, 2015).

3. Daniel Sargent, "Pax Americana: Sketches for an Undiplomatic History," *Diplomatic History* 42 (2018): 357–376. On the long history of US intervention in international health, see Randall Packard, *A History of Global Health: Interventions into the Lives of Other Peoples* (Baltimore: Johns Hopkins University Press, 2016).

4. Daniel Rodgers, *Age of Fracture* (Cambridge, MA: Harvard University Press, 2011).

5. Jay Sexton, "From Triumph to Crisis: An American Tradition" (Stuart L. Bernath Lecture, Society for Historians of American Foreign Relations Annual Luncheon, American Historical Association Annual Meeting, January 5, 2019). See also his book *A Nation Forged by Crisis: A New American History* (New York: Basic Books, 2018).

6. Madeleine Bunting, "Small Is Beautiful—An Economic Idea That Has Sadly Been Forgotten," *The Guardian*, November 10, 2011.

7. Andrew Simms, "Small Is Beautiful . . . but Schumacher's Economics of Scale Runs Deeper," *The Guardian*, November 14, 2011.

8. On the history of US imperialism during this period, see Daniel Immerwahr, *How to Hide an Empire: A History of the Greater United States* (New York: MacMillan, 2019).

9. Aimé Césaire, *Discourse on Colonialism*, 2nd ed. (Dakar: Présence Africaine, 1955); Albert Memmi, *The Colonizer and the Colonized* (1974; repr., London: Earthscan Publications, 2010); Frantz Fanon, *The Wretched of the Earth* (New York: Grove Press, 1963), originally published in French as *Les damnés de la Terre* (1961).

10. Abena Osseo-Asare, "Scientific Equity: Experiments in Laboratory Education in Ghana," *Isis* 104 (December 2013): 713–741.

11. Samuel Moyn, *The Last Utopia: Human Rights in History* (Cambridge, MA: Harvard University Press, 2012).

12. Samuel Moyn, *Not Enough: Human Rights in an Unequal World* (Cambridge, MA: Belknap Press of Harvard University Press, 2018), 146.

13. Achille Mbembe, *On the Postcolony* (Berkeley: University of California Press, 2001), originally published in 2000 in French as *De la postcolonie: Essai sur l'imagination politique dans l'Afrique contemporaine.*

14. Surendra Chopra, "The Emerging Trends in the Non-Aligned Movement," *Indian Journal of Political Science* 47 (1986): 161–177.

15. David Ekbladh, *The Great American Mission: Modernization and the Construction of an American World Order* (Princeton, NJ: Princeton University Press, 2009), 226–227; USAID, "Foreign Aid Explorer: The Official Record of US Foreign Aid" (data set), https://explorer.usaid.gov/aid-trends.html. For 1960s USAID budget figures in nominal dollars, see the USAID "Greenbook": *US Overseas Loans and Grants and Assistance from International Organizations: Obligations and Loan Authorizations, July 1, 1945–June 30, 1969*, Special Report Prepared for the House Foreign Affairs Committee, April 24, 1970, USAID Development Experience Clearinghouse. For historical overall budget numbers, see *The Budget of the United States Government: Fiscal Year 1969* (Washington, DC: US Government Printing Office, 1968), https://fraser.stlouisfed.org/title/54/item/19022/toc/380798.

16. Constant dollars are adjusted for inflation. Constant dollar figures available from https://explorer.usaid.gov/aid-trends.html. USAID budget figures from *US Overseas Loans and Grants and Assistance from International Organizations: Obligations and Loan Authorizations, July 1, 1945-June 30, 1974*, 1975, USAID Development Experience Clearinghouse. For historical overall budget numbers, see *The Budget of the United States Government: Fiscal Year 1974* (Washington, DC: US Government Printing Office, 1974), https://fraser.stlouisfed.org/files/docs/publications/usbudget/bus_1974.pdf.

17. Heidi Morefield, "More with Less: Commerce, Technology, and International Health at USAID, 1961–1981," *Diplomatic History* 43, no. 4 (2019): 618–643.

18. Carroll Pursell, "Technology and Social Inequality," *Spontaneous Generations: A Journal for the History and Philosophy of Science* 8 (2016): 22–26.

19. Michel Foucault, in his series of lectures at the Collège de France in 1979 (unpublished until 2004), *The Birth of Biopolitics*, traces a longer history for the "neoliberal state" that began in the interwar period with the Freiburg school in Germany and the Chicago school in the United States. See Michel Foucault, *The Birth of Biopolitics: Lectures at the Collège de France, 1978–79*, ed. Michael Senellart (Basingstoke, UK: Palgrave Macmillan, 2008). In the mid-1970s and 1980s, experiments with the "neoliberal state" extended to specific economic restructuring in countries like Chile and Argentina.

Historians of capitalism have similarly reconsidered the boundaries between liberal and conservative economic policies in recent years. The historian Angus Burgin, for example, resituates the origins of neoliberal policy in the Great Depression of the 1930s with economist Friedrich Hayek and the network of scholars that went on to be formalized as the Mont Pèlerin Society in 1947 but shows that these policy strands were continually reinvented throughout the twentieth century by economists and politicians. See Angus Burgin, *The Great Persuasion: Reinventing Free Markets Since the Depression* (Cambridge, MA: Harvard University Press, 2012), 8–10, 120–121. The historian Ben Jackson has likewise shown that the neoliberalism of the 1930s and 1940s was not, as is commonly believed, against the welfare state but rather saw its primary target as socialist central planning and embraced many aspects of state expansion in support of the free market while admonishing earlier *laissez faire* market liberals. For Jackson and for Burgin, pre-1980s neoliberalism was an ideology in flux. As a glib shorthand for the turn toward Reaganomics, therefore, neoliberalism does not necessarily signify what authors who use the term in the context of late twentieth century global health intend. See Ben Jackson "The Origins of Neo-Liberalism: The Free Economy and the Strong State, 1930–1947," *Historical Journal* 53, 1 (2010): 129–151, and Ben Jackson, "Hayek, Keynes, and the Origins of Neo-Liberalism: A Reply to Farrant and McPhail," *Historical Journal* 55, no. 3 (2012): 779–783.

20. Carroll Pursell, "The Rise and Fall of the Appropriate Technology Movement in the United States, 1965–1985," *Technology and Culture* 34, no. 2 (July 1993): 629–637; Carroll Pursell, "The Hoe or the Tractor? Appropriate Technology and American Technical Aid after World War II," *Icon* 5 (1999): 90–99.

21. Langdon Winner, "Building the Better Mouse Trap," in *The Whale and the Reactor: A Search for Limits in an Age of High Technology* (Chicago: University of Chicago Press, 1986), 80.

22. On the agency of actors in the Global South, see Gabrielle Hecht, introduction to *Entangled Geographies: Empire and Technopolitics in the Global Cold War*, ed. Gabrielle Hecht (Cambridge, MA: MIT Press, 2011); Gabrielle Hecht, *Being Nuclear: Africans and the Global Uranium Trade* (Cambridge, MA: MIT Press, 2012); Suzanne Moon "Takeoff or Self-Sufficiency? Ideologies of Development in Indonesia, 1957–1961," *Technology and Culture* 39, no. 2 (1998): 187–212; Itty Abraham, *The Making of the Indian Atomic Bomb: Science, Secrecy, and the Postcolonial State* (London: Zed Books, 1998). On critiques of technological determinism, see Merritt Roe Smith, "Technological Determinism in American Culture," in *Does Technology Drive History?*, ed. Merritt Roe Smith and Leo Marx (Cambridge, MA: MIT Press, 1994); Langdon Winner, "Do Artifacts Have Politics?" in *The Social Shaping of Technology*, ed. Donald MacKenzie and Judy Wajcman, 2nd ed. (Maidenhead, UK: Open University Press, 1999).

23. See, for example, Nancy Leys Stepan, *Eradication: Ridding the World of Diseases Forever?* (Ithaca, NY: Cornell University Press, 2011); Matthew Connelly, *Fatal Misconception: The Struggle to Control World Population* (Cambridge, MA: Harvard University Press, 2008); Randall Packard, *The Making of a Tropical Disease: A Short History of Malaria* (Baltimore: Johns Hopkins University Press, 2007).

24. Andrew Russell and Lee Vinsel, "After Innovation, Turn to Maintenance," *Technology & Culture* 59 (2018): 1–25.

25. David Edgerton, *The Shock of the Old: Technology and Global History since 1900* (Oxford: Oxford University Press, 2006); David Arnold, *Everyday Technology: Machines and the Making of India's Modernity* (Chicago: University of Chicago Press, 2013).

26. Clapperton Mavhunga *Transient Workspaces: Technologies of Everyday Innovation in Zimbabwe* (Cambridge, MA: MIT Press, 2014), 7–8.

27. See Frederick Cooper and Randall Packard, *International Development and the Social Sciences: Essays on the History and Politics of Knowledge* (Berkeley: University of California Press, 1997); James Ferguson, *The Anti-Politics Machine: "Development," Depoliticization, and Bureaucratic Power in Lesotho* (Cambridge: Cambridge University Press, 1990); David Mosse, *Cultivating Development: An Ethnography of Aid Policy and Practice* (London: Pluto Press, 2005); Judith Justice, *Policy, Plans, and People: Foreign Aid and Health Development* (Berkeley: University of California Press, 1986).

28. Important exceptions are Jeremy Greene's work tracing the history of the essential drugs concept, Peter Redfield's work on medical humanitarian kits, and Alice Street's ongoing work on diagnostic devices. See Jeremy A. Greene, "Making Medicines Essential: The Emergent Centrality of Pharmaceuticals in Global Health," *BioSo-*

cieties 6, no. 1 (2011): 10–23; Peter Redfield, *Life in Crisis: The Ethical Journey of Doctors without Borders* (Berkeley: University of California Press, 2013); Alice Street, "The Testing Revolution: Investigating Diagnostic Devices in Global Health," *Somatosphere*, April 9, 2018, http://somatosphere.net/2018/04/testing-revolution.html.

29. See, for example, Marcos Cueto, "The Origins of Primary Health Care and Selective Primary Health Care," *American Journal of Public Health* 94, no. 11 (2004), 1864–1874. On the World Bank, see Salmaan Keshavjee, *Blind Spot: How Neoliberalism Infiltrated Global Health* (Berkeley: University of California Press, 2014), 95–96, 103, 106.

30. There are, of course, notable exceptions. See Cooper and Packard, *International Development.*

31. Peter Redfield, "Fluid Technologies: The Bush Pump, the LifeStraw and Microworlds of Humanitarian Design," *Social Studies of Science* 46, no. 2 (2016): 159–183.

32. On communal water pumps, see Marianne de Laet and Annemarie Mol, "The Zimbabwe bush pump: Mechanics of a fluid technology," *Social Studies of Science* 30, no. 2 (2000): 225–263. On this genre of profitable social entrepreneurship for appropriate technology, see Raphael Kaplinsky, "Schumacher Meets Schumpeter: Appropriate Technology below the Radar," *Research Policy* 40 (2011): 193–203. On the shifting discourse of human rights from the collective to the individual, see Moyn, *Not Enough.*

33. For example, see Abena Osseo-Asare, *Bitter Roots: The Search for Healing Plants in Africa* (Chicago: University of Chicago Press, 2014); Timothy Burke, *Lifebuoy Men, Lux Women: Commodification, Consumption, and Cleanliness in Modern Zimbabwe* (Durham, NC: Duke University Press, 1996); Kristin Peterson, *Speculative Markets: Drug Circuits and Derivative Live in Nigeria* (Durham: Duke University Press, 2014); Nicholas Thomas, *Entangled Objects: Exchange, Material Culture, and Colonialism in the Pacific* (Cambridge, MA: Harvard University Press, 1991).

CHAPTER ONE

1. Rangoon is now known as Yangon. It was the Burmese capital until 2002, when the capital was moved to Naypyitaw. For more on the history of Prome Court, see "June XI Business Centre," Architectural Guide Yangon, https://www.yangongui.de/june-xi-business-centre/. The Pegu Club is a Victorian Gentleman's club, opened in 1871, which originated the gin-based cocktail by the same name. The teak structure was built in 1882. Its membership was limited to white people throughout the British colonial era. For more on the Pegu Club, see "Pegu Club," Architectural Guide Yangon, https://www.yangongui.de/pegu-club/.

2. United Nations, "Letter of Appointment: to Mr. E. F. Schumacher," Project Personnel, July 27, 1954, box 1, folder 1, Schumacher Personal Papers, Schumacher Center for New Economics, Great Barrington, MA.

3. E. F. Schumacher to U Thant, March 29, 1955, Rangoon, box 1, folder 5, Schumacher Personal Papers.

4. Schumacher to U Thant, March 29, 1955, Schumacher Personal Papers.

5. Schumacher to U Thant, March 29, 1955, Schumacher Personal Papers.

6. Alden Whitman, "U Thant Is Dead of Cancer at 65," *New York Times*, November 26, 1974, 1.

7. E. F. Schumacher to A. J. Wakefield, March 11, 1955, Rangoon, box 1, folder 5, Schumacher Personal Papers.

8. "Farewell to Mr. E. F. Schumacher," invitation from U Thant to E. F. Schumacher, March 30, 1955, Rangoon, box 1, folder 1, Schumacher Personal Papers.

9. See S. Habib Ahmed to E. F. Schumacher, confidential letter, February 9, 1956, box 1, folder 5, Schumacher Personal Papers.

10. Robert Leonard, "Schumacher and 'Buddhist Economics,'" *Journal of the History of Economic Thought* 41, no. 2 (2019): 159–186.

11. For more on the history of modernization theory, see Nils Gilman, *Mandarins of the Future: Modernization Theory in Cold War America* (Baltimore: Johns Hopkins University Press, 2003).

12. As quoted in the biography written by his daughter, Barbara Wood, *Alias Papa: A Life of Fritz Schumacher* (London: Butler & Tanner, 1984), 244.

13. Wood, 245.

14. Leonard, "Schumacher and 'Buddhist Economics.'"

15. Wood, *Alias Papa*, 245; see also Leonard, "Schumacher and 'Buddhist Economics.'"

16. "The Threefold Refuge and the Eight Precepts," undated note, box 1, folder 1, in Schumacher Personal Papers.

17. Kirkpatrick Sale, *Human Scale: Revisited* (White River Junction, VT: Chelsea Green Publishing, 2017), 131–133.

18. Robert R. Nathan & Associates for the Burmese Economic and Social Board, *Pyidawtha—The New Burma* (London: Hazell Watson and Viney, 1954), 83–105, 110.

19. Robert R. Nathan & Associates, *Pyidawtha*, 28.

20. For more on the way Third World leaders used their position between the United States and the Soviets, see Odd Arne Westad, *The Global Cold War* (Cambridge: University of Cambridge Press, 2007).

21. Robert J. McMahon, Harriet D. Schwar, and Louis J. Smith, eds., *Foreign Relations of the United States, 1955–1957, Southeast Asia*, vol. 22, *Burma* (Washington, DC: Government Printing Office, 1989), doc. 20.

22. Wood, *Alias Papa*, 248.

23. E. F. Schumacher interview with Kalpana Sharma, September 3, 1977, box 3, folder 1, Schumacher Personal Papers. The story is also recounted in a footnote in Leonard, "Schumacher and 'Buddhist Economics.'"

24. E. F. Schumacher, "Economics in a Buddhist Country," draft paper, 1955, box 5, folder 6, Schumacher Personal Papers. The essay was later published: E. F. Schumacher, "Economics in a Buddhist Country," appendix A in in Jayaprakash Narayan's *A Plea for Reconstruction of Indian Polity*, published by Secretary Akhil Bharat Sarva Seva Sangh Prakashan, Wardha (Bombay State) (Varanasi, India: B. B. Press, 1959), 108–117.

25. Narayan, *Plea*.

26. Quotes from Narayan, *Plea*. The Robert R. Nathan Associates plan was developed between August 1951 and August 1953 with the American engineering firm Knappen Tippets Abbett McCarthy and the management consulting firm Pierce Management. It comprised two volumes. Both are available at https://www.nathaninc.com/insight/economic-and-engineering-development-of-burma-1953/.

27. Schumacher, "Economics in a Buddhist Country" (1959).

28. Schumacher.

29. Schumacher.

30. Schumacher.

31. P. A. V. Spencer to E. F. Schumacher, January 11, 1955, Rangoon, box 1, folder 1, Schumacher Personal Papers.

32. Spencer to Schumacher, January 11, 1955, Schumacher Personal Papers.

33. E. F. Schumacher, "Buddhist Economics," in *Small Is Beautiful: Economics as If People Mattered* (London: Blond and Briggs, 1973), ch. 4.

34. Schumacher, 53.

35. See Leonard, "Schumacher and 'Buddhist Economics,'" 161; Wood, *Alias Papa*, 107–114.

36. David Astor's parents, Lord and Lady Astor, owned a successful business empire as well as the *Observer* newspaper.

37. Wood, *Alias Papa*, 132.

38. E. F. Schumacher, "Multilateral Clearing," *Economica* 10, no. 38 (May 1943): 150–165.

39. Wood, *Alias Papa*, 133–134. Schumacher was, at the time, very happy to have someone take up the idea, even though he did not receive explicit credit. In his later years, Schumacher would accuse Keynes of plagiarism—many passages from Keynes's paper bear a distinct resemblance to Schumacher's own words.

40. Wood, *Alias Papa*, 135. See also Charles H. Hession, "Schumacher as Heir to Keynes' Mantle," *Review of Social Economy* 44, no. 1 (April 1986): 1–12. Sir Richard W. B. Clarke, known as Otto, was a British statistician and civil servant.

41. E. F. Schumacher, "Obituary: Lord Keynes," *The Times*, April 22, 1946, 7.

42. Wood, *Alias Papa*, 139–140.

43. Unfortunately, while there is a trove of official documents and formal papers at the Schumacher archives, there is little in the way of personal explication or self-reflection.

44. Wood, *Alias Papa*, 256.

45. Charles Fager, "Small Is Beautiful, and So Is Rome: Surprising Faith of E. F. Schumacher," *Christian Century*, April 6, 1977, 325.

46. Wood, *Alias Papa*, 259–261.

47. Wood, 260.

48. From E. F. Schumacher, lecture 5 at London University, October 28, 1959, as quoted in Wood, *Alias Papa*, 260.

49. From E. F. Schumacher, lecture to Imperial College, 1965, as quoted in Wood, *Alias Papa*, 261.

50. Wood, *Alias Papa*, 262.

51. Wood, 265.

52. E. F. Schumacher, lecture 17 at London University, 1960, as quoted in Wood, 265.

53. Wood, 266.

54. On the entry of economic calculus into development thinking, and the entwining of the concept of economy with population, see Michelle Murphy, *The Economization of Life* (Durham, NC: Duke University Press, 2017).

55. Timothy Mitchell, *Rule of Experts: Egypt, Techno-Politics, Modernity* (Berkeley: University of California Press, 2002), 51.

56. This can be read as a somewhat ironic form of relativism, given his criticism of Einstein.

57. E. F. Schumacher, "Establishing a New Level of Technology," *The Times*, August 14, 1965, xiv.

58. E. F. Schumacher, "Non-Violent Economics," *Observer*, August 21, 1960.

59. Daniel Immerwahr, *Thinking Small: The United States and the Lure of Community Development* (Cambridge, MA: Harvard University Press, 2015), 71.

60. For more on the differences between Nehru and Gandhi's development philosophies, particularly regarding technology, see David Arnold, *Everyday Technology: Machines and the Making of India's Modernity* (Chicago: University of Chicago Press, 2013).

61. For a detailed outline of Gandhian economics and how this informs the appropriate technology concept, see Ram Swarup, *Gandhian Economics: A Supporting Technology* (Lucknow, India: Appropriate Technology Development Association, 1977); M. M. Hoda, "India's Experience and the Gandhian Tradition," in *Appropriate Technology: Problems and Promises*, ed. Nicolas Jéquier (Paris: Development Centre of the Organisation for Economic Co-operation and Development [OECD], 1976), 144–155. For more on Gandhian economics more generally, see Kenneth Rivett, "The Economic Thought of Mahatma Gandhi," *British Journal of Sociology* 10, no. 1 (March 1959): 1–15.

62. Immerwahr, *Thinking Small*, 70.

63. Immerwahr, 70.

64. The essay was later republished as a chapter in *Small Is Beautiful: Economics as If People Mattered* (London: Blond and Briggs, 1973).

65. J. C. Kumarappa, *Economy of Permanence: A Quest for a Social Order Based on Non-Violence* (Varanasi, India: Sarva Seva Sangh Prakashan, 1945), https://www.mkgandhi.org/ebks/economy-of-permanence.pdf.

66. Kumarappa, 13.

67. M. K. Gandhi, *Harijan* 64 (September 1, 1937): 217–218, in *Gandhiji on Villages*, ed. Divya Joshi (Mumbai: Gandhi Book Centre, 2002), 21.

68. For more on the history of peasant studies and how it affected international development, see Mitchell, *Rule of Experts*.

69. James C. Scott, *Seeing Like a State: How Certain Schemes to Improve the Human Condition Have Failed* (New Haven, CT: Yale University Press, 1998), 131, 114. Le Corbusier was invited to take over the project after Mayer's partner, the architect Matthew Nowicki, died in a plane crash in 1950.

70. Immerwahr, *Thinking Small*, 71–72.

71. Immerwahr, 75.

72. Nicole Sackley, "Village Models: Etawah, India, and the Making and Remaking of Development in the Early Cold War," *Diplomatic History* 37, no. 4 (2013): 759.

73. Nick Cullather, *The Hungry World: America's Cold War Battle against Poverty in Asia* (Cambridge, MA: Harvard University Press, 2010), 77; David Engerman, *The Price of Aid: The Economic Cold War in India* (Cambridge, MA: Harvard University Press, 2018), 70.

74. Robert R. Nathan Associates, *Pyidawtha*, 160.

75. Robert R. Nathan Associates, 160.

76. Robert R. Nathan Associates, 160.

77. See M. M. Coady, *Masters of Their Own Destiny: The Story of the Antigonish Movement of Adult Education through Economic Cooperation* (New York: Harper & Row Publishers, 1939).

78. Chris Armstrong, "Economics after God's Own Image," *Christian History* 75 (2002).

79. Jane Jacobs, *The Death and Life of Great American Cities* (New York: Random House, 1961). See Also James C. Scott, *Seeing Like a State*, 132–146.

80. Schumacher, "Establishing a New Level," xiv

81. E. F. Schumacher, "How to Help Them Help Themselves," *The Observer*, August 29, 1965, 17.

82. Schumacher, "How to Help Them." On the history of community development in the United States, see Immerwahr, *Thinking Small.*

83. Witold Rybczynski, *Paper Heroes: A Review of Appropriate Technology* (New York: Anchor Books / Doubleday, 1980).

84. Malcolm Hollick, "The Appropriate Technology Movement and Its Literature: A Retrospective," *Technology in* Society 4, no. 3 (1982): 213–229. See also Anthony Akubue, "Appropriate Technology for Socioeconomic Development in Third World Countries," *Journal of Technology Studies* 26, no. 1 (2000): 33–43.

85. U. Pellegrini, "The Problem of Appropriate Technology," *IFAC Proceedings Volumes* 12, no. 6 (September 1979): 1–5.

86. E. F. Schumacher, *Small Is Beautiful: Economics as If People Mattered* (London: Blond and Briggs, 1973; New York: Harper Perennial, 2010), 198.

CHAPTER TWO

1. E. F. Schumacher, *Small Is Beautiful: Economics as If People Mattered* (London: Blond and Briggs, 1973; New York: Harper Perennial, 2010), 146.

2. George Orwell would later work for *The Observer*, the London paper that David Astor edited.

3. Kohr, Schumacher, and many of their fellows are what Claus-Dieter Krohn has called "intellectuals in exile"—refugee scholars who fled to the United States from the Nazi regime. Krohn's work shows they had a particularly large influence in the founding of the New School for Social Research. See Claus-Dieter Krohn, *Intellectuals in Exile: Refugee Scholars and the New School for Social Research* (Amherst: University of Massachusetts Press, 1993), originally published by Campus-Verlag in 1987 as *Wissenschaft im Exil.* Historian Udi Greenberg has likewise looked at the influence of refugee scholars on United States policy during the early Cold War years: *The Weimar Century: German Émigrés and the Ideological Foundations of the Cold War* (Princeton, NJ: Princeton University Press, 2014).

4. Conversation with Susan Witt, executive director of the Center for New Economics, Great Barrington, MA, December 11, 2017. See also Associated Press, "Dr. Leopold Kohr, 84; Backed Smaller States," *New York Times*, February 28, 1994, D9.

5. Leopold Kohr, "Why Small Is Beautiful: The Size Interpretation of History" (Ninth Annual E. F. Schumacher Lectures, Mount Holyoke College, South Hadley, MA, October 1989), ed. Hildegarde Hannum.

6. Ivan Illich, "The Wisdom of Leopold Kohr" (Fourteenth Annual E. F. Schumacher Lectures, October 1994, Yale University, New Haven, CT), ed. Hildegarde Hannum.

7. Associated Press, "Dr. Leopold Kohr, 84."

8. Illich, "Wisdom of Leopold Kohr," 1994.

9. Kirkpatrick Sale, foreword to Leopold Kohr, *The Breakdown of Nations*, rev. ed. (New York: Dutton, 1978).

10. Leopold Kohr, "Disunion Now," in *The Breakdown of Nation* (London: Routledge and Kegan Paul, 1957), ch. 3.

11. Kohr.

12. Kohr.

13. Chase Madar, "The People's Priest," *American Conservative*. February 1, 2010, http://www.theamericanconservative.com/articles/the-peoples-priest/.

14. Author interview with Tom Dewar, October 10, 2016.

15. Francine du Plessix Gray, "Profiles: Rules of the Game," *New Yorker*, April 25, 1970, 49.

16. Dewar interview, October 10, 2016.

17. On development, see Ivan Illich, "To Hell with Good Intentions" (address to the Conference on InterAmerican Student Projects, Cuernavaca, Mexico, April 20, 1968), http://www.swaraj.org/illich_hell.htm. On medicine, see Ivan Illich, *Medical Nemesis: The Expropriation of Health* (London: Calder and Boyars, 1975).

18. For more on the role of expertise in development, see Timothy Mitchell, *Rule of Experts: Egypt, Techno-Politics, Modernity* (Berkeley: University of California Press, 2002).

19. Dewar interview, October 10, 2016.

20. Dewar interview, October 10, 2016.

21. Jerry Brown, "Ivan Illich," in "Remembering Ivan Illich—Reflections on a Seminal Cultural Critic/Intellectual Gadfly," *Whole Earth Review* (Spring 2003).

22. Jesse McKinley, "How Jerry Brown Became 'Governor Moonbeam,'" *New York Times*, March 6, 2010; Reid Wilson, "From Moonbeam to Mainstream: Jerry Brown in Winter," *The Hill*, January 24, 2018, http://thehill.com/homenews/state-watch/370374-governor-moonbeam-sees-his-ideas-become-mainstream.

23. Christopher Cadelago, "Jerry Brown Takes International Stage on Nuclear Danger: 'We All Ought to Wake Up,'" *Sacramento Bee*, March 20, 2017.

24. Brown, "Ivan Illich." See also Jerry Brown, "Letter to the Editor re: The Life of Ivan Illich," *New York Times*, December 11, 2002.

25. Theodore Roszak, introduction to E. F. Schumacher, *Small Is Beautiful: Economics as If People Mattered*, 2nd ed. (New York: Harper and Row, 1975), 3.

26. Roszak, introduction to Schumacher, *Small Is Beautiful*, 5.

27. Theodore Roszak, *The Making of a Counter Culture: Reflections on a Technocratic Society and Its Youthful Oppositions* (Berkeley: University of California Press, 1969).

28. Roszak, introduction to Schumacher, *Small Is Beautiful*, 5.

29. Schumacher, *Small Is Beautiful*, 271.

30. Roszak, introduction to Schumacher, *Small Is Beautiful*, 4.

31. Roszak, 4.

32. He wrote that "economics, which Lord Keynes had hoped would settle down as a modest occupation similar to dentistry, suddenly [became] the most important subject of all." Schumacher, *Small Is Beautiful*, 69.

33. Schumacher, 79. See also M. M. Coady, *Masters of Their Own Destiny: The Story of the Antigonish Movement of Adult Education through Economic Cooperation* (New York: Harper & Row Publishers, 1939).

34. Schumacher, *Small Is Beautiful*, 164.

35. Schumacher, 204.

36. Schumacher, 204.

37. Schumacher, 208.

38. Schumacher, 209. Note that while Schumacher typically continued to use the phrase "intermediate technology" in his writing, even after the 1968 conference, he frequently used the term "appropriate" to describe the technology or "gifts" that development programs would distribute, as in this quote.

39. Schumacher, 209.

40. Heidi Morefield, "More with Less: Commerce, Technology, and International Health at USAID 1961–1981," *Diplomatic History* 43, no. 4 (2019): 618–643.

41. Schumacher, *Small Is Beautiful*, 146.

42. Schumacher, 186.

43. Schumacher, 187.

44. In the journal *Science*, Nicholas Wade wrote that "perhaps the chief lacuna of *Small Is Beautiful* is that it describes a number of maybe utopian ideals without offering many signposts as to how they may be attained." Nicholas Wade, "E. F. Schumacher: Cutting Technology Down to Size," *Science* 189, no. 4198 (July 1975): 199–201.

45. Robert Swann, "E. F. Schumacher—Small Is Beautiful," in *Peace, Civil Rights, and the Search for Community: An Autobiography*, February 1998, https://centerforneweconomics.org/publications/peace-civil-rights-and-the-search-for-community-an-autobiography/#Chapter%2025.

46. "The Hundred Most Influential Books since the War," *Times Literary Supplement*, October 6, 1995, https://www.the-tls.co.uk/articles/private/the-hundred-most-influential-books-since-the-war/.

47. Arghiri Emmanuel, *Appropriate or Underdeveloped Technology?* (New York: John Wiley & Sons, 1982). Emmanuel lived and worked in the Congo from 1937 to 1957, apart from a few years during World War II. He supported the Congolese independence movement and Patrice Lumumba, but he left the country as the political situation deteriorated. His arguments also frequently invoke dependency theory, which was a critical response to the dominant modernization theory the West propounded through foreign aid.

48. Emmanuel, 48. He references G. L. Reuber, *Le Rôle des investissements privés* (Paris: OECD, 1974), as the source of the description.

49. Samir Amin, *Imperialism and Unequal Development* (New York: Monthly Review Press, 1977), 172.

50. Amin, 173.

51. Schumacher, *Small Is Beautiful*, 181.

52. Robert Vitalis, "The Midnight Ride of Kwame Nkrumah and Other Fables of Bandung," *Humanity* (Summer 2013): 261–288.

53. Vitalis, 262. See also Peter Willetts, *The Non-Aligned Movement: The Origins of a Third World Alliance* (London: Pinter, 1978); Itty Abraham, "From Bandung to NAM: Non-Alignment and Indian Foreign Policy, 1947–65," *Commonwealth and Comparative Studies* 46 (2009): 195–219.

54. Vitalis, "Midnight Ride," 273.

55. U Thant, *View from the UN* (Garden City, NY: Doubleday, 1978), 20. For a full analysis of Buddhist precepts as applied to U Thant's record as secretary-general, see Walter Dorn, "U Thant: Buddhism in Action," in *The UN Secretary-General and Moral Authority: Ethics and Religion in International Leadership*, ed. Kent Kille (Washington, DC: George Washington University Press, 2007), 143–186.

56. David Spanier, "Governor Brown Lives Up to His Image: Politician with a Cool, Fresh Mind Who Sounds Like an Aspiring PhD in Sociology," *The Times*, December 1, 1977, 6; William Tuohy, "Brown Makes Small Talk in London: Parries with Press, Eulogizes Economist Schumacher," *Los Angeles Times*, December 1, 1977, B29.

57. "Community Health Care—Schumacher Style," *The Lancet*, November 26, 1977, 1114–1115.

58. "Community Health Care," 1114.

59. "Community Health Care," 1114.

60. E. F. Schumacher, "People's Power" (address at the annual general meeting of the National Council of Social Service, London, December 1974).

CHAPTER THREE

1. "Dale B. Fritz" (obituary), *Schenectady (NY) Daily Gazette*, February 17, 2013, https://www.legacy.com/obituaries/dailygazette/obituary.aspx?n=dale-b-fritz&pid=163095954.

2. VITA is still around today, however it is now known as EnterpriseWorks/VITA. Although a comprehensive history of international volunteerism in this period is lacking, the movement and spirit of the VITA volunteers coincided with the formation of the Peace Corps. See Elizabeth Cobbs Hoffman, *All You Need Is Love: The Peace Corps and the Spirit of the 1960s* (Cambridge, MA: Harvard University Press, 1998).

3. Daniel Immerwahr, *Thinking Small: The United States and the Lure of Community Development* (Cambridge, MA: Harvard University Press, 2015), 55.

4. Volunteers in Technical Assistance (VITA), *The Village Technology Handbook*, 2nd ed. (Washington, DC: US Agency for International Development, Communication Resources Department, 1965), 3.

5. VITA, 3.

6. Eric Hobsbawm, "The Machine Breakers" *Past & Present* 1 (1952): 57–70.

7. Hobsbawm, 59–60.

8. Pushing back on the many laudatory studies of innovation as the primary form of progress, there has been a recent turn toward scholarship on maintenance in science and technology studies (STS) and the history of technology. See, for example, Andrew Russell and Lee Vinsel, "After Innovation, Turn to Maintenance," *Technology & Culture* 59 (2018): 1–25.

9. Although there are "treatment" sections for both malaria and bilharziasis, the handbook does not offer any easy or accessible solutions. Rather, it directs readers that both diseases require drugs that only "well-trained persons" should administer. See VITA, *Village Technology Handbook* (Arlington, VA: Volunteers in Technical Assistance, 1988). This publication was the third major revision. The health and sanitation section of the first edition, in 1963, focused exclusively on latrine construction. A second printing with minor revisions expanded the section in 1965. The second edition in 1970 added information on bilharziasis. The handbook remained more or less unchanged through printings in 1975 and 1978. Malaria was added in the third edition, in 1988. All editions and printings were funded by USAID.

10. See VITA, *Village Technology Handbook* (Schenectady, NY: Volunteers in Technical Assistance, 1970), v. The language of mastery and of the role of community development in state building is reminiscent of Rev. Moses Coady's writings for the Antigonish movement and the Coady International Institute, which was established just a few years before VITA published the first edition of the handbook.

11. It was in the same spirit that both USAID and the Peace Corps enthusiastically translated and distributed copies of David Werner's 1970 handbook *Donde no hay doctor*, a simple, easy-to-follow guide to treating all manner of health problems with simple tools and remedies, intended for village health workers. See David Werner, *Where There Is No Doctor: A Village Health Care Handbook* (Berkeley, CA: Hesperian Health Guides, 1970).

12. "News from VITA," *Peace Corps Volunteer*, December 1962, 21, https://peace corpsonline.org.

13. "Survival Switchboard," *Peace Corps Volunteer*, May–June 1970, 23–24. The plans for the laundry machine are still popular online today on Christian survivalist websites.

14. Hon. Daniel E. Button, "VITA: An Inspiring Example of Responsible and Constructive Cooperation between the Government and the Private Sector," *Congressional Record: Proceedings and Debates of the 90th Congress, Second Session* 114, pt. 4 (February 26, 1968): 4186–4187.

15. Button.

16. Button.

17. See Heidi Morefield, "More with Less: Commerce, Technology, and International Health at USAID 1961–1981," *Diplomatic History* 43, no. 4 (2019): 618–643.

18. Intermediate Technology Development Group (ITDG), *ITDG Bulletin*, no. 3 (August 1968): 5, in unprocessed box 15, Intermediate Technology Development Group Archives, Schumacher Center for New Economics, Great Barrington, MA (hereafter, ITDG Archives).

19. ITDG, 5.

20. This tendency to privilege designers over workers and technicians has been thoroughly critiqued in the literature of the history of technology and in science and technology studies. See, for example, Steven Shapin, "The Invisible Technician," *American Scientist* 77 (1989): 554–563.

21. ITDG, *ITDG Bulletin*, no. 3 (August 1968): 6, in ITDG Archives.

22. The CINVA-RAM block press was designed by engineer Raúl Ramírez at the Centro Interamericano de Vivienda y Planeamiento (CINVA) in Bogotá in 1957, through US foreign aid programs in self-help housing. See Amy Offner, *Sorting Out the Mixed Economy: The Rise and Fall of Welfare and Developmental States in the Americas* (Princeton, NJ: Princeton University Press, 2019), 86–87.

23. ITDG, *ITDG Bulletin*, no. 3 (August 1968): 6, in boxes 6–7, ITDG Archives.

24. Office of the War on Hunger, USAID, "Training Center," in *War on Hunger: A Report from the Agency for International Development* 1, no. 1 (January 1968): 20.

25. Stewart Brand, *The Whole Earth Catalog*, Fall 1968, 18.

26. See "Dr. Schumacher on Foreign Aid," *The Times*, October 24, 1966, 13. ITDG is still active today but has changed its name to Practical Action and moved to a small town, Bourton-on-Dunsmore, near Rugby, in the United Kingdom.

27. ITDG, *ITDG Bulletin*, no. 1 (September 1966): 13, in ITDG Archives.

28. ITDG, *ITDG Bulletin*, no. 2 (March 1968): 6, in ITDG Archives.

29. ITDG, 3.

30. ITDG, 5.

31. ITDG, 4.

32. The University of Science and Technology was known after Ghanaian independence up until the military coup of 1966 as the Kwame Nkrumah University of Science and Technology, named after the independence leader and first president of Ghana. That original name was restored in 1998.

33. ITDG, *ITDG Bulletin*, no. 3 (August 1968): 3, quoting *Kumasi Pioneer*, June 1, 1968, in ITDG Archives.

34. ITDG, 4.

35. See George McRobie and Sir John Palmer, "A Report on the Establishment of a Technology Consultancy Centre at the University of Science and Technology, Kumasi,

Ghana," ITDG, January 11, 1971, UNIDO Open Archives, https://open.unido.org/api/documents/4683793/download/GHANA.%20A%20REPORT%20ON%20THE%20ESTABLISHMENT%20OF%20A%20TECHNOLOGY%20CONSULTANCY%20CENTRE%20AT%20THE%20UNIVERSITY%20OF%20SCIENCE%20AND%20TECHNOLOGY%20AT%20KUMASI%20(972.en).

36. Professor S. Sey, "Proposal for a Technology Advisory Centre at the University of Science and Technology, Kumasi," appendix 4, p. 3, in *Report on the Panel of Experts on Advanced Institutes for Applied Science and Technology in Africa*, UN Economic Commission for Africa session at the University of Manchester, August 10–14, 1970, http://repository.uneca.org/bitstream/handle/10855/9942/Bib-50563.pdf?sequence=1.

37. UN Economic Commission for Africa, "Report of the Secretariat on the Work of the Commission, 15 February, 1970–31 January, 1971," 14, http://repository.uneca.org/bitstream/handle/10855/3813/Bib-29708.pdf?sequence=1.

38. See the website of the Technology Consultancy Centre, at https://tcc.knust.edu.gh.

39. "Low-Cost Hospital Equipment at Ahmadu Bello Hospital," *ITDG Bulletin*, no. 4 (February 1969): 5–8, in ITDG Archives.

40. Manuals and books in unprocessed box 5, ITDG Archives.

41. George McRobie, *Small Is Possible* (London: Jonathan Cape, 1981).

42. Stewart Brand, "Photography Changes Our Relationship to Our Planet," Smithsonian Photography Initiative, May 30, 2008, https://web.archive.org/web/20080530221651/http://click.si.edu/Story.aspx?story=31.

43. Brand.

44. Brand.

45. Thomas Rid, *Rise of the Machines* (New York: W. W. Norton & Co., 2016), 167–168.

46. W. Patrick McCray, *The Visioneers: How a Group of Elite Scientists Pursued Space Colonies, Nanotechnologies, and a Limitless Future* (Princeton, NJ: Princeton University Press, 2013), 204.

47. Andrew Kirk, *Counterculture Green: The Whole Earth Catalog and American Environmentalism* (Lawrence: University of Kansas Press, 2007), 1.

48. The historian Margaret O'Mara also credits Schumacher with influencing the counterculture in the computing industry. See O'Mara, *The Code: Silicon Valley and the Remaking of America* (New York: Penguin Press, 2019), 125–126.

49. Brand became very interested in systems thinking and cybernetics at Stanford. For more on his early life, see Fred Turner, *From Counterculture to Cyberculture: Steward Brand, the Whole Earth Network, and the Rise of Digital Utopianism* (Chicago: University of Chicago Press, 2006), 43.

50. Stewart Brand, "Low Impact Technology Group," *Whole Earth Epilog*, September 1974, 534.

51. Stewart Brand, "Planet Economics," *Whole Earth Epilog*, September 1974, 466.

52. Turner, *From Counterculture to Cyberculture*. See also See O'Mara, *The Code: Silicon Valley and the Remaking of America* (New York: Penguin Press, 2019).

53. Kirk, *Counterculture Green*.

54. Stewart Brand, *Whole Earth Epilog*, September 1974, 451, 608–612.

55. Brand, 582, 737.

56. Stewart Brand, "We Are as Gods," *Whole Earth Catalog*, Fall 1968, 2.

57. Stewart Brand, "We Are as Gods," Understanding Whole Systems, http://www.wholeearth.com/issue/1010/article/195/we.are.as.gods.

58. Edmund Leach, *A Runaway World?* (New York: Oxford University Press, 1968).

59. During the 1960s and 1970s, Brand identified as a libertarian. For more on his thinking on politics and climate change, see McCray, *Visioneers*.

60. Brand, "We Are As Gods," Understanding Whole Systems. On how this attitude has played out in the twenty-first century, see Moira Wegel, "Silicon Valley's Sixty-Year Love Affair with the Word 'Tool,'" *New Yorker*, April 11, 2018.

61. Turner, *From Counterculture to Cyberculture*, 42–43.

62. Anna Wiener, "The Complicated Legacy of Stewart Brand's 'Whole Earth Catalog,'" *New Yorker*, November 16, 2018.

63. Wiener.

64. Of course, Brand's repudiation of libertarianism does not stop modern libertarians from holding him up as a hero. For example, *Reason* magazine, which espouses "free minds, free markets," published a very celebratory fifty-year commemoration of Brand and *Whole Earth* in December 2018, including Brand's (the catalog's) recommendations for Ayn Rand's *Atlas Shrugged* and Milton Friedman's *Capitalism and Freedom*. See Nick Gillespie, "We Are as Gods and Might as Well Get Good at It," *Reason*, December 2018, https://reason.com/2018/11/04/we-are-as-gods-and-might-as-we.

65. Wiener, "Complicated Legacy."

66. Susan Rennie and Kirsten Grimstad, *The New Woman's Survival Catalog* (New York: Coward, McCann & Geoghegan, 1973), 7.

67. Rennie and Grimstad, 7.

68. Rennie and Grimstad, 7.

69. Second-wave feminists are frequently critiqued for their lack of attention to the plight of women of color, who face the double burden of being marginalized for their sex and for their race. Advanced first by Kimberlé Crenshaw, the analytical concept of intersectionality posits that the result of the intersections of these two forms of discrimination is greater than the sum of its parts. See Kimberlé Crenshaw, "Demarginalizing the Intersection of Race and Sex: A Black Feminist Critique of Antidiscrimination Doctrine, Feminist Theory and Antiracist Politics," *University of Chicago Legal Forum* 1989, no. 1 (1989): 139–168. For another germinal work critiquing second-wave feminist theorists like Betty Friedan for the exclusion of women of color, lower-class women, and queer women, see bell hooks, *Feminist Theory: From Margin to Center* (Boston: South End Press, 1984). There have been recent attempts to "reread" and reinterpret second-wave feminists, partly on the basis of the assertion that they were unfairly maligned in the Reagan 1980s and that those unfair characterizations have dominated the second wave's historical treatment. See, for example, the 2014 essay collection in the *New Statesman*, including Helen Lewis, "Rereading the Second Wave: Why Feminism Needs to Respect Its Elders," *New Statesman*, May 12, 2014, https://www.newstatesman.com/helen-lewis/2014/05/rereading-second-wave-why-feminism-needs-respect-its-elders. See also Kirsten Swinth, *Feminism's Forgotten Fight: The Unfinished Struggle for Work and Family* (Cambridge, MA: Harvard University Press, 2018).

70. Rennie and Grimstad, *New Woman's Survival Sourcebook*, 232.

71. See Daniel Patrick Moynihan, *The Negro Family: The Case for National Action* (Washington, DC: US Department of Labor, Office of Policy and Planning Research, March 1965), https://web.archive.org/web/20140428151006/http://www.dol.gov/

dol/aboutdol/history/webid-meynihan.htm. For a historical analysis of the Moynihan Report's legacy, see Daniel Geary, *Beyond Civil Rights: The Moynihan Report and Its Legacy* (Philadelphia: University of Pennsylvania Press, 2015).

72. Moynihan, *Negro Family*, vii.

73. Moynihan, 236.

74. Radicalesbians, *The Woman-Identified Woman* (pamphlet), 1970, Atlanta Lesbian Feminist Alliance Archives, Duke University, https://repository.duke.edu/dc/wlmpc/wlmms01011. The manifesto was written in response to then president of the National Organization for Women Betty Friedan's 1969 expulsion of lesbian women from the organization, calling them a "lavender menace." The Radicalesbians' manifesto is widely seen as a founding document in lesbian feminism.

75. For an overview of the long history of feminist movements' entanglements with health care, see Susan Reverby, "Feminism & Health," *Health and History* 4 (2002): 5–19.

76. On *Our Bodies, Ourselves*, see Wendy Kline, *Bodies of Knowledge: Sexuality, Reproduction, and Women's Health in the Second Wave* (Chicago: University of Chicago Press, 2010).

77. Rennie and Grimstad, *New Woman's Survival Sourcebook*, 71–92. For more on the history of vaginal self-exams as a feminist reconceptualization of medical care and reproduction, see Michelle Murphy, *Seizing the Means of Reproduction: Entanglements of Feminism, Health, and Technoscience* (Durham, NC: Duke University Press, 2012).

78. Kevin Hjortshøj O'Rourke, Ahmed Rahman, and Alan Taylor, "Luddites, the Industrial Revolution, and the Demographic Transition," *Journal of Economic Growth* 18 (2013): 374.

79. See Mar Hicks, *Programmed Inequality: How Britain Discarded Women Technologists and Lost Its Edge in Computing* (Cambridge, MA: MIT Press, 2017).

80. Meg Miller, "Behind the Making of the 'Feminist *Whole Earth Catalog*': It's the Pre-Internet Feminist Internet," *Eye on Design*, August 16, 2018, https://eyeondesign.aiga.org/behind-the-making-of-the-feminist-whole-earth-catalog/.

81. Miller.

82. Carole Levine, "Women," *Whole Earth Epilog*, September 1974, 582.

83. Levine.

84. See also Morefield, "More with Less."

85. The historian Daniel Rodgers refers to the last quarter of the twentieth century as the "age of fracture." See Daniel Rodgers, *Age of Fracture* (Cambridge, MA: Belknap Press of Harvard University Press, 2011).

CHAPTER FOUR

1. Robert Swann, "E. F. Schumacher—Small Is Beautiful," in *Peace, Civil Rights, and the Search for Community: An Autobiography*, February 1998, https://centerforneweconomics.org/publications/peace-civil-rights-and-the-search-for-community-an-autobiography/#Chapter%2025.

2. "The Daily Diary of President Jimmy Carter: March 22, 1977," White House, https://www.jimmycarterlibrary.gov/assets/documents/diary/1977/d032277t.pdf.

3. On the balance of bilateral versus multilateral aid spending in the United States, see Lee Milton Howard, untitled paper beginning "Trends in US and International Financial Support," box 3, folder 3.13, Lee Milton Howard Collection, Alan Mason

Chesney Medical Archives of the Johns Hopkins Medical Institutions, Baltimore; Lee Milton Howard, "A Profile of US Development Cooperation in Health Sectors," July 1981, box 3, folder 3.58, Lee Milton Howard Collection; "Aid Architecture," World Bank, International Development Association Resource Mobilization, http://documents.worldbank.org/curated/en/745221468313781790/Aid-architecture-an-overview-of-the-main-trends-in-official-development-assistance-flows; Homi Kharas, "Rethinking the Role of Multilaterals in the Global Aid Architecture," 2010 Brookings Blum Roundtable Policy Briefs, https://www.brookings.edu/wp-content/uploads/2016/07/09_development_aid_kharas2–2.pdf.

4. Rainer's legacy is controversial—he began his architectural career in 1939 as a member of the Nazi Party and worked in service of the Nazi regime throughout the war, although he remains a celebrated hero of Austrian architecture in the country.

5. Theodore Hesburgh, "Address by Ambassador Theodore M. Hesburgh, Chairman, US Delegation to the United Nations Conference on Science and Technology for Development" (Vienna, August 20, 1979), 1, Digital Collections of University of Notre Dame, https://archives.nd.edu/Hesburgh/UDIS-H2-12-06.pdf.

6. Hesburgh, 2.

7. US Government, "Public Law 93-189, December 17, 1973," sec. 2 (2)(B)(b)(3), http://uscode.house.gov/statutes/pl/93/189.pdf.

8. For a detailed account of this history, see Heidi Morefield, "More with Less: Commerce, Technology, and International Health at USAID, 1961–1981," *Diplomatic History* 43, no. 4 (2019): 618–643.

9. Private and Voluntary Organizations (PVOs) were an antecedent category to nongovernmental organizations (NGOs). The former grouped together for- and nonprofit organizations, while the latter is typically understood to be reserved for nonprofit entities.

10. On the Peace Corps involvement, see Michael O'Brien, *Hesburgh: A Biography* (Washington, DC: Catholic University of America Press, 1998), 86–89. On his work in the civil rights movement, see Paul T. Murray, "'To Change the Face of America': Father Theodore M. Hesburgh and the Civil Rights Commission," *Indiana Magazine of History* 111 (June 2015): 121–154.

11. Theodore Hesburgh, "Statement by Commissioner Hesburgh," *US Commission on Civil Rights Report, Book 5: Justice* (Washington, DC, 1961), 167–168, https://www.law.umaryland.edu/marshall/usccr/documents/cr11961bk5.pdf.

12. Michael Dennis argues that this delay in establishing a coordinated science and technology policy was due to the specter of Lysenkoism and scientists' fears of government intervention and control. See Michael Aaron Dennis, "Our Monsters, Ourselves: Reimagining the Problem of Knowledge in Cold War America," in *Dreamscapes of Modernity: Sociotechnical Imaginaries and the Fabrication of Power*, ed. Sheila Jasanoff and Sang-Hyun Kim (Chicago: University of Chicago Press, 2015).

13. The year 1955 was also the year that twenty-nine newly independent states in Africa and Asia, including India, Indonesia, Gold Coast, and Egypt, got together at the Bandung Conference, which focused on opposing colonialism and neocolonialism by both the West and the Soviets. In 1961, these same countries formed the Non-Aligned Movement at the Belgrade Conference, which had a similar mandate to walk a "middle path" between the West and the Soviet Union. Though not explicitly linked, Hammarskjöld's decision to host the UNCAST conference should be considered in the context of this shifting political landscape.

14. W. Murray Todd to Guyford Stever, October 25, 1978, box 199: Post-Government Subject File, folder: UN Conference on Science and Technology for Development—Background (3), H. Guyford Stever Papers, 1930–1990, Gerald Ford Presidential Library, Ann Arbor, MI.

15. Dr. Walsh McDermott, "The 1963 UN Conference on the Application of Science and Technology (UNCAST) for the Benefit of the Less Developed Countries," 1–7, September 25, 1978, box 47, folder 6, Walsh McDermott Personal Papers, Weill-Cornell Medical Center Archives, New York.

16. For more on Cold War science and the competition between the United States and Soviet Union, see Audra Wolfe, *Competing with the Soviets: Science, Technology, and the State in Cold War America* (Baltimore: Johns Hopkins University Press, 2013).

17. Millikan led the Center for International Studies at MIT, an interdisciplinary research group that included Walt W. Rostow and developed modernization theory. See Nils Gilman, *Mandarins of the Future: Modernization Theory in Cold War America* (Baltimore: Johns Hopkins University Press, 2003).

18. McDermott, "1963 UN Conference," 17.

19. McDermott, 18.

20. For more on the politics of postcolonial efforts to equalize the distribution of material resources, see Samuel Moyn, *Not Enough: Human Rights in an Unequal World* (Cambridge, MA: Belknap Press of Harvard University Press, 2018), 100–118.

21. Jean M. Wilkowski, *Abroad for Her Country: Tales of a Pioneer Woman Ambassador in the US Foreign Service* (Notre Dame, IN: University of Notre Dame Press, 2008), 14.

22. "We got along famously," Wilkowski was quoted as saying of her relationship with Father Hesburgh. O'Brien, *Hesburgh*, 149.

23. Rose Bannigan (National Academy of Sciences, Commission on International Relations) to Participants, Discussion Seminar on Regional Science and Technology Development in the Middle East, Re: Meeting, Tuesday October 10, 1978, memorandum, September 20, 1978, and "Discussion Seminar on Regional Science and Technology Development in the Middle East," October 10, 1978, both in box 127: Post-Government Subject File, folder: BOSTID-Discussion Seminar on Regional Science and Technology Development in the Middle East (1), H. Guyford Stever Papers.

24. Bannigan to Participants, 3–5.

25. Michael Oppenheimer and John Pothier, "The Role of Technology in North/South Relations," box 199: Post-Government Subject File, folder: United Nations Conference on Science and Technology for Development Background (5), H. Guyford Stever Papers.

26. Oppenheimer and Pothier, 4.

27. Oppenheimer and Pothier, 10.

28. "Overview," 5–9, in US Delegation, *United Nations Conference on Science and Technology for Development, Briefing Book*, box 200: Post-Government Subject File, folder: UN Conference on Science and Technology for Development—Briefing Book (1), H. Guyford Stever Papers.

29. "Overview," 6.

30. "Minutes of Meeting, US Delegation for UNCSTD," Office of the US Coordinator for the UN Conference on Science and Technology for Development, Department of State, June 7, 1979, box 200: Post-Government Subject File, folder: UN Conference on Science and Technology for Development—General (1) 9/78–8/79, H. Guyford Stever Papers.

31. "Minutes of Meeting," 1.

32. The United Nations, through the UN Industrial Development Organization (UNIDO), also held a series of meetings on appropriate technology and published monographs for different sectors in preparation for UNCSTD. One of these, *Appropriate Industrial Technology for Drugs and Pharmaceuticals*, grew out of a meeting of the Second Consultative Group on Appropriate Industrial Technology, which met in June 1978 in Vienna. It contained a highly technical action plan for the development of pharmaceutical industries in developing countries. See UNIDO, *Appropriate Industrial Technology for Drugs and Pharmaceuticals*, Monographs on Appropriate Industrial Technology 10 (New York: United Nations, 1980).

33. James H. Scheuer, "Opening Statement," in "Appropriate Technology: Hearings before the Subcommittee on Domestic and International Scientific Planning, Analysis, and Cooperation of the Committee on Science and Technology," US House of Representatives, 95th Cong., 2nd sess., July 25–27, 1978, no. 110, p. 3.

34. US Delegation, "Code of Conduct Relating to Transnational Corporations," in *UNCSTD Briefing Book*, box 200: Post-Government Subject File, folder: UN Conference on Science and Technology for Development—Briefing Book (3), H. Guyford Stever Papers.

35. US Delegation, 3.

36. Gustav Ranis, "Appropriate Technology: Obstacles and Opportunities," in "Appropriate Technology: Hearings before the Subcommittee on Domestic and International Scientific Planning, Analysis, and Cooperation of the Committee on Science and Technology," US House of Representatives, 95th Cong., 2nd sess., July 25–27, 1978, no. 110, p. 55.

37. Ranis, 66.

38. Ranis, 77.

39. Sander Levin, "Statement before the Subcommittee on Domestic and International Scientific Planning, Analysis, and Cooperation, July 27, 1978," in "Appropriate Technology: Hearings before the Subcommittee on Domestic and International Scientific Planning, Analysis, and Cooperation of the Committee on Science and Technology," US House of Representatives, 95th Cong., 2nd sess., July 25–27, 1978, no. 110, p. 332.

40. Levin, 334, 336.

41. Levin, 336–337.

42. See "The United Nations Conference on Science and Technology for Development and an AID Program of Assistance to Developing Countries for Conference Preparations" in Henry A. Arnold to Robert D. Simpson, Re: Small Activity—Assistance to Mauritania in Preparation for UN Conference, May 3, 1978, Development Experience Clearinghouse, USAID, https://dec.usaid.gov/dec/content/Detail_Presto.aspx?vID=47&ctID=ODVhZjk4NWQtM2YyMi00YjRmLTkxNjktZTcxMjM2NDBmY2Uy&rID=MjYyMTc%3d.

43. See "United Nations Conference on Science and Technology for Development and an AID Program."

44. Without Mauritanian records on this subject, it is not possible to make any definitive claims; nevertheless, a few well-informed speculations can be made on the basis of the secondary literature.

45. Gregory Mann, *From Empires to NGOs in the West African Sahel: The Road to Nongovernmentality* (Cambridge: Cambridge University Press, 2015), 141.

46. Mann, 142.

47. Mann, 172.

48. Mann. Charles Piot, too, remarks about the increasing phenomenon of NGOs replacing the state in governance in West Africa, in his case in Northern Togo. However, Piot periodizes this shift later, to the early 1990s. See Charles Piot, *Nostalgia for the Future: West Africa after the Cold War* (Chicago: University of Chicago Press, 2010), esp. ch. 5, "Arrested Development."

49. "The United Nations Conference on Science and Technology for Development and an AID Program of Assistance to Developing Countries for Conference Preparations," 29–30, in Henry A. Arnold to Robert D. Simpson, Re: Small Activity—Assistance to Mauritania in Preparation for UN Conference, May 3, 1978, Development Experience Clearinghouse, USAID.

50. International Science and Technology Institute, "Papers on Science and Technology Policy, Industry and Agriculture for the Government of the Islamic Republic of Mauritania in Preparation for the 1979 United Nations Conference on Science and Technology for Development (UNCSTD)," 2, July 15, 1978, Development Experience Clearinghouse, USAID.

51. International Science and Technology Institute, 3 (original emphasis).

52. International Science and Technology Institute, 5.

53. International Science and Technology Institute, 18–19. This is a markedly different approach from the one Tamara Giles-Vernick describes USAID contributing to in the Central African Republic in the 1980s, wherein it helped establish a long-term conservation zone specifically for research purposes. See Giles-Vernick, *Cutting the Vines of the Past: Environmental Histories of the Central African Rain Forest* (Charlottesville: University of Virginia Press, 2002), 44.

54. International Science and Technology Institute, "Papers on Science and Technology Policy, Industry and Agriculture," 20.

55. International Science and Technology Institute, 72.

56. International Science and Technology Institute, 78. For more on model villages, see chapters 1 and 5.

57. The military regime from July 1978 to the end of 1979, led by Col. Mustafa Ould Salek and the Comité Militaire de Redressement National junta remained close to France. By early 1980, a second military coup by Col. Mohamed Khouna Ould Haidallah had strained relations with France as the new Comité Militaire de Salut National regime sought to pull out of the conflict in Western Sahara (thus abandoning France's ally Algeria in the fight).

58. For more on the shifting trends in international and global health—between narrow, vertical campaigns and horizontal, broad-based ones—see Randall Packard, *A History of Global Health: Interventions into the Lives of Other Peoples* (Baltimore: Johns Hopkins University Press, 2016).

59. Edward Kennedy, "Keynote Address: Pharmaceuticals for Developing Countries," in *Pharmaceuticals for Developing Countries: Conference Proceedings* (Washington, DC: National Academies Press, 1979), 5. On essential drugs as appropriate technologies, see also Chung-Hae Ahn, "After Alma-Ata: A Conceptual Agenda," December 22, 1978, box 3, folder 3.30, Lee Milton Howard Collection.

60. Kenneth Warren, "Panel Presentation," in *Pharmaceuticals for Developing Countries: Conference Proceedings* (Washington, DC: National Academies Press, 1979), 413.

61. For a critique of selective primary health care as essentially a defense of tradi-

tional vertical approaches to health systems, see Oscar Gish, "Selective Primary Health Care: Old Wine in New Bottles," *Social Sciences of Medicine* 16 (1982): 1049–1054.

62. Although McKeown had published his initial findings in a series of four articles in *Population Studies* between 1955 and 1972, McDermott references the following two synthesis papers: Thomas McKeown, "Historical Perspective on Science and Health" (paper presented at the annual meeting of the Institute of Medicine, National Academy of Sciences, Washington, DC, October 1976); and McKeown, *Role of Medicine: Dream, Mirage, or Nemesis* (London, Nuffield Provincial Hospitals Trust, 1976). For more on the historical impact of the McKeown thesis, see James Colgrove, "The McKeown Thesis: A Historical Controversy and Its Enduring Influence," *American Journal of Public Health* 92, no. 5 (2002): 725–729.

63. Walsh McDermott, "Historical Perspective," in *Pharmaceuticals for Developing Countries: Conference Proceedings* (Washington, DC: National Academies Press, 1979), 16.

64. Gilbert Omenn, "International Health: From the Perspective of the Carter Administration," in *Pharmaceuticals for Developing Countries: Conference Proceedings* (Washington, DC: National Academies Press, 1979), 171–172.

65. See *Pharmaceuticals for Developing Countries: Conference Proceedings* (Washington, DC: National Academies Press, 1979), 255, 405, 409.

66. Developing World Industry and Technology, "Changes in the Terms and Conditions of Technology Transfer by the Pharmaceutical Industry to Newly Industrializing Nations over the past Decade," July 1979, 3 (prepared for USAID, under contract AID/DSAN-147–697, order no. 3698077), Development Experience Clearinghouse, USAID.

67. See Jack Baranson, *North-South Technology Transfer* (Mt. Airy, MD: Lomond Publications, 1981); Jack Baranson, *Technology and the Multinationals: Corporate Strategies in a Changing World Economy* (Lexington, MA: Lexington Books, 1978); Jack Baranson, "Automotive Industries in Developing Countries" (World Bank Staff Occasional Paper No. 8), International Bank for Reconstruction and Development, 1969; Jack Baranson, *Technology for Underdeveloped Areas: An Annotated Bibliography* (Oxford, UK: Pergamon Press, 1967). Baranson was also a panelist, with Yale development economist Gustav Ranis, at one of the UN Advisory Committee on the Application of Science and Technology to Development (ACAST) symposia, "Science and Technology in Development Planning," held in Mexico City, May 28–June 1, 1979, in preparation for UNCSTD. See Victor Urquidi, ed., *Science and Technology in Development Planning: Science, Technology and Global Problems* (Oxford, UK: Pergamon Press, 1979).

68. For more on pharmaceutical programs for developing countries during and after the failure of the Vienna conference at the United Nations and the World Health Organization, including the Action Program on Essential Drugs, see Jeremy Greene, "Pharmaceutical Geographies: Mapping the Boundaries of the Therapeutic Revolution" in *Therapeutic Revolutions: Pharmaceuticals and Social Change in the Twentieth Century*, ed. Jeremy Greene, Flurin Condrau, and Elizabeth Siegel Watkins (Chicago: University of Chicago Press, 2016).

69. Heidi Morefield, oral history interview with Donald A. Henderson, January 6, 2016, Baltimore.

70. Institute of Medicine, Division of International Health, "Review of the Agency for International Development Health Strategy," 5, September 1978, National Academy of Sciences, Washington, DC, box 199: Post-Government Subject File, folder: UN Con-

ference on Science and Technology for Development—Background (2), H. Guyford Stever Papers, quoting Walsh McDermott, "Modern Medicine and the Demographic-Disease Pattern of Overly Traditional Societies: A Technologic Misfit," *Journal of Medical Education* 41, no. 137 (1966).

71. For more on the debate over primary health care versus targeted disease campaigns, see Randall Packard, *A History of Global Health: Interventions into the Lives of Other Peoples* (Baltimore: Johns Hopkins University Press, 2016).

72. Institute of Medicine, "Review of the Agency for International Development Health Strategy," 10.

CHAPTER FIVE

1. How appropriate technology caught on within different organizations can be productively theorized with two different models: on an intellectual level, appropriate technology seemed to be an innovation that was compatible with many existing organizational values and beliefs and therefore diffused and was adopted by a network of leaders engaged in international public health; on a managerial level, appropriate technology could be conceived of as a "fad"—"the strong inclinations of contemporary management to seize on novel approaches with great enthusiasm, and often to dismiss the solution a few years later in favor of another new solution." On the former, see Everett M. Rogers, *Diffusion of Innovations*, 5th ed. (New York: Simon & Schuster, 2003). On the latter, see Margaret Brindle and Peter Stearns, *Facing Up to Management Faddism: A New Look at an Old Force* (Westport, CT: Quorum Books, 2001), vii.

2. See Dr. Julia Walsh and Dr. Kenneth Warren, "Selective Primary Health Care: An Interim Strategy for Disease Control in Developing Countries," *New England Journal of Medicine* 308, no. 18 (November 1, 1979): 967–974. This paper was originally presented at the Bellagio, Italy, conference in April 1979.

3. Socrates Litsios, "The Long and Difficult Road to Alma-Ata: A Personal Reflection," *International Journal of Health Services* 32, no. 4 (2002): 709–732. Marcos Cueto, "The Origins of Primary Health Care and Selective Primary Health Care," *American Journal of Public Health* 94, no. 11 (2004): 1864–1874; Theodore M. Brown, Marcos Cueto, and Elizabeth Fee, "The World Health Organization and the Transition from 'International' to 'Global' Public Health," *American Journal of Public Health* 96, no. 1 (2006): 62–72; Socrates Litsios, *The Third Ten Years of the World Health Organization, 1968–1977* (Geneva: World Health Organization, 2008).

4. Nitsan Chorev, *The World Health Organization between North and South* (Ithaca, NY: Cornell University Press, 2012).

5. Quinn Slobodian, *Globalists: The End of Empire and the Birth of Neoliberalism* (Cambridge, MA: Harvard University Press, 2018).

6. For more on how appropriate technology was defined in the developing world, see chapters 1 and 5.

7. Adom Getachew, *Worldmaking after Empire: The Rise and Fall of Self-Determination* (Princeton, NJ: Princeton University Press, 2019), 144–145.

8. UN General Assembly, Resolution 3201 (S-VI) "Declaration on the Establishment of a New International Economic Order," May 1, 1974, http://www.un-documents.net/s6r3201.htm.

9. "Rene Maheu, 70, of UNESCO, Is Dead," *New York Times*, December 20, 1975, 30.

10. Céline Giton, "Weapons of Mass Distribution: UNESCO and the Impact of Books," in *A History of UNESCO: Global Actions and Impacts*, ed. Poul Duedahl (New York: Palgrave Macmillan, 2016), 49–72.

11. Maheu's career at UNESCO, in various executive roles, lasted from 1946 to 1974. He died in 1975.

12. Phillip W. Jones, "René Maheu and UNESCO's 'Decisive Mutation,'" in *International Policies for Third World Education: UNESCO, Literacy, and Development* (London: Routledge, 1988).

13. UNESCO, *Science and Technology in African Development*, Science Policy Studies and Documents no. 35: Report on CASTAFRICA (Paris: UNESCO Press, 1974), 41.

14. UNESCO, 42.

15. Merritt E. Kimball to William K. Gamble, "Inter-Office Memorandum Re: CASTAFRICA Conference, Dakar, 21–30 January, 1974," February 5, 1974, Reports 011989, Ford Foundation Records, Rockefeller Archive Center.

16. Kimball to Gamble, 43.

17. Senghor was so beloved by the Senegalese people that they supported him in fending off an attempted 1963 coup d'état. He was the first African president to leave office voluntarily, in 1980.

18. Kimball to Gamble, "Inter-Office Memorandum," 43.

19. On novel fictions and reimaginations of future social structures built around racial equity, see Ruha Benjamin, "Racial Fictions, Biological Facts: Expanding the Sociological Imagination through Speculative Methods," *Catalyst: Feminism, Theory, Technoscience* 2, no. 2 (2016): 1–28. Wakanda was created by Marvel Comic's Stan Lee and Jack Kirby in 1966, first appearing in the comic *Fantastic Four*. Arcadia was a common theme in European Renaissance art, depicted as a pastoral utopia of harmonious nature.

20. UNESCO, *Science and Technology*, 135–136.

21. UNESCO,134.

22. UNESCO, 50.

23. See Getachew, *Worldmaking after Empire*.

24. Audra Wolfe, *Competing with the Soviets: Science, Technology, and the State in Cold War America* (Baltimore: Johns Hopkins University Press, 2012).

25. Associated Press, "Britain Following Lead of US, Will Withdraw from UNESCO," *Los Angeles Times*, December 5, 1985.

26. Jo Thomas, "Britain Confirms Its Plan to Quit a 'Harmfully Politicized' UNESCO," *New York Times*, December 6, 1985.

27. Litsios, *Third Ten Years*, 20.

28. Litsios, *Third Ten Years*.

29. Ted Brown, Elizabeth Fee, and Victoria Stepanova, "Halfdan Mahler: Architect and Defender of the World Health Organization 'Health for All by 2000' Declaration of 1978," *American Journal of Public Health* 106, no. 1 (January 2016): 38–39.

30. Randall Packard, *A History of Global Health: Interventions into the Lives of Other Peoples* (Baltimore: Johns Hopkins University Press, 2016), 89–90.

31. Anil Agarwal, "Mahler's Revolutionary Study," *Nature* 284 (March 20, 1980): 206–209.

32. For more on Schumacher's relationship with Keynes, see chapter 1.

33. Samuel Moyn, *Not Enough: Human Rights in an Unequal World* (Cambridge, MA: Belknap Press of Harvard University Press, 2018), 146–147.

34. See S. R. Sen to Mr. Robert McNamara, "Office Memorandum," April 9, 1971, folder 1772424, Robert S. McNamara Personal Chronological Files—Chrons 14, World Bank Archives, Washington, DC.

35. Sen to McNamara. See also E. F. Schumacher, "Intermediate Technology: The Missing Factor in Foreign Aid," *Oxford Diocesan Magazine*, July 1970.

36. World Bank, "Appropriate Technology in World Bank Activities" (Report Number UNN405), July 19, 1976, i, World Bank Archives.

37. World Bank, 36.

38. See Jeremy Greene, "Vital Objects: Essential Drugs and Their Critical Legacies," in *Reimagining (Bio) Medicalization, Pharmaceuticals, and Genetics: Old Critiques and New Engagements*, ed. Susan Bell and Anne Figert (New York: Routledge, 2015).

39. Halfdan Mahler, "The Work of WHO 1975" (address to the Twenty-Ninth World Health Assembly, May 4, 1976, Geneva), in *Official Records of the World Health Organization No. 229*, viii.

40. Mahler, xiii–xiv.

41. "WHO Chief Criticizes Costly Health Technology and Urges Replacement," *New York Times*, May 3, 1977, 13.

42. World Health Organization, "Global Strategy for Health for All by the Year 2000" (Geneva: WHO, 1981); Packard, *History of Global Health*, 227.

43. See, for example, Randall Packard, *Global Health: A History of Interventions into the Health of Other Peoples* (Baltimore: Johns Hopkins University Press, 2016); Nitsan Chorev, *The World Health Organization between North and South* (Ithaca, NY: Cornell University Press, 2012); Socrates Litsios, "The Long and Difficult Road to Alma-Ata: A Personal Reflection," *International Journal of Health Services* 32, no. 4 (2002): 709–732; Marcos Cueto, "The Origins of Primary Health Care and Selective Primary Health Care," *American Journal of Public Health* 94, no. 11 (2004): 1864–1874.

44. Halfdan Mahler, "The Meaning of Health for All by the Year 2000," *World Health Forum* 2, no. 1 (1981): 5–22. On the debates over defining essential drugs, see Jeremy Greene, "Making Medicines Essential: The Emergent Centrality of Pharmaceuticals in Global Health," *BioSocieties* 6, no. 1 (2011): 10–33.

45. World Health Organization, "Report of the Inter-Regional Workshop on Appropriate Technology for Health, New Delhi: 24–31 March, 1981," N85-445-3, Jacket 8, p. 3, WHO Archives.

46. For more on the ITDG, see chapter 2.

47. John W. Forje, *The Rape of Africa at Vienna: African Participation in the 1979 United Nations Conference on Science and Technology for Development* (Lund, Sweden: Action-Oriented Research Group on African Development, 1979). See also Forje's earlier volume: John W. Forje, Abdallah Makange, and Chamberlain Nii-Djan France, *The Challenge of African Economic and Technological Development and the Influence of International Forums* (Lund, Sweden: Action-Oriented Research Group on African Development, 1978).

48. Forje, *Rape of Africa at Vienna*, 58.

49. Forje, 67.

50. Forje, 61, 65.

51. "'Arrested' Retort" *Retort*, no. 3, August 22, 1979, 1, box 200: Post-Government Subject File, folder: UN Conference on Science and Technology for Development—Coverage in Newsletters, H. Guyford Stever Papers, 1930–1990, Gerald Ford Presidential Library, Ann Arbor, MI.

52. "We're Open to Offers Say US," *Retort*, no. 2, August 21, 1979, 1, box 200: Post-Government Subject File, folder: UN Conference on Science and Technology for Development—Coverage in Newsletters, H. Guyford Stever Papers.

53. "We're Open."

54. "STOP THE CONFERENCE!—We Want to Get Off," *Retort*, no. 4, August 23, 1979, 1, box 200: Post-Government Subject File, folder: UN Conference on Science and Technology for Development—Coverage in Newsletters, H. Guyford Stever Papers.

55. "STOP THE CONFERENCE!"

56. Peter Stalker, "Fund OK—But Now How Big?" *Retort*, no. 9, August 30, 1979, 1, 8, box 200: Post-Government Subject File, folder: UN Conference on Science and Technology for Development—Coverage in Newsletters, H. Guyford Stever Papers.

57. Stalker, 1.

58. Martin M. Kaplan, "A Personal Reflection," *Bulletin of the Atomic Scientists*, December 1979, 53–55.

59. Jean Wilkowski, *Abroad for Her Country: Tales of a Pioneer Woman Ambassador in the US Foreign Service* (Notre Dame, IN: University of Notre Dame Press, 2008), 317. For more on VITA, see chapter 2.

60. The "US National Paper" (their position paper for the conference) was not without contemporary critics. Ward Morehouse of the Council on International and Public Affairs and a political scientist at Columbia University, who participated in the American Association for the Advancement of Science preparatory workshops, wrote that the National Paper was "disappointing" and did not go far enough to "change the rules of the game," for example with international patent laws. See "US Report Fails to Bridget Gap with Third World," *Nature*, 277, February 22, 1979, 593.

61. Executive Order 12163—Administration of Foreign Assistance and Related Function, September 29, 1979, http://www.archives.gov/federal-register/codification/executive-order/12163.html.

62. "Statement of Thomas Ehrlich, Director, US IDCA before the Committee on Foreign Affairs, House of Representatives," February 5, 1980, 18, Development Experience Clearinghouse, USAID.

63. Editor, "A Modest Aid Proposal," *Washington Post*, July 7, 1979, A12; Daniel S. Greenberg, "A Bum Rap on the New Research Institute," *Washington Post*, November 13, 1979, A19.

64. Dennis DeConcini, "Bureaucratic Bonanza," *Washington Post*, December 4, 1979, A21.

65. Werner Fornos, "Just Another Research Subsidy," *Washington Post*, July 24, 1979, A13.

66. See Douglas Bennet to Tom Ehrlich, "Responsibility for Science and Technology" (memo), August 13, 1979, box 12, folder 3, Douglas Bennet Papers, Special Collections & Archives, Wesleyan University, Middletown, CT; Douglas Bennet to Frank Press, "AID Science and Technology Advisor" (memo), April 9, 1980, box 13, folder 5, Douglas Bennet Papers.

67. See the US Delegation UNCSTD Briefing Book, "The UNCSTD Program of Action: Problem," box 200: Post-Government Subject File, folder: UN Conference on Science and Technology for Development—Briefing Book (2), H. Guyford Stever Papers.

68. Recent literature on the ethics of using indigenous peoples and bodies as re-

search subjects has called into question the common practice in global health of trialing new drugs and devices in the Global South. See Adriana Petryna, *When Experiments Travel: Clinical Trials and the Global Search for Human Subjects* (Princeton, NJ: Princeton University Press, 2009); Melissa Graboyes, *The Experiment Must Continue: Medical Research and Ethics in East Africa, 1940–2014* (Athens: University of Ohio Press, 2015); Johanna Tayloe Crane, *Scrambling for Africa: AIDS, Expertise, and the Rise of American Global Health Science* (Ithaca, NY: Cornell University Press, 2013); Christopher R. Henke, "Making a Place for Science: The Field Trial," *Social Studies of Science* 30, no. 4 (June 29, 2016).

69. Technologies deemed "appropriate" could also be theorized as "immutable mobiles." See Bruno Latour, *Science in Action: How to Follow Scientists and Engineers through Society* (Cambridge, MA: Harvard University Press, 1987). For critiques of this approach to international health and development, see Frederick Cooper and Randall Packard, *International Development and the Social Sciences: Essays on the History and Politics of Knowledge* (Berkeley, CA: University of California Press, 1997); James Ferguson, *The Anti-Politics Machine: "Development," Depoliticization, and Bureaucratic Power in Lesotho* (Cambridge: Cambridge University Press, 1990); David Mosse, *Cultivating Development: An Ethnography of Aid Policy and Practice* (London: Pluto Press, 2005); Sunil Amrith, *Decolonizing International Health: India and Southeast Asia, 1930–65* (New York: Palgrave Macmillan, 2006); Judith Justice, *Policy, Plans, and People: Foreign Aid and Health Development* (Berkeley, CA: University of California Press, 1986).

70. NGO Forum: Science and Technology for Development (pamphlet and program), August 19–29, 1979, Series X: Miscellaneous UNCSTD and UNESCO Materials, box 14, folder: NGO Forum American Association for the Advancement of Science Archives: Science, Technology, and Development Records, Washington, DC.

CHAPTER SIX

1. Bill Gates interview with *BBC Newsround*, December 7, 2001, http://news.bbc.co.uk/cbbcnews/hi/club/your_reports/newsid_1697000/1697132.stm.

2. Stewart Brand, *The Media Lab: Inventing the Future at MIT* (New York: Viking Adult, 1987), 22.

3. "PATH: 40 Years of Innovation and Impact," PATH, February 6, 2017, https://www.path.org/articles/path-40-years-of-innovation-and-impact/.

4. Heidi Morefield interview with Richard Mahoney, February 14, 2018.

5. "PATH: 40 Years," quote derived from Jolayne Houtz (PATH) interview with Richard Mahoney, November 2015, PATH private internal archive.

6. "PATH: 40 Years."

7. Heidi Morefield, "More with Less: Commerce, Technology, and International Health at USAID, 1961–1981," *Diplomatic History* 43, no. 4 (2019): 618–643.

8. Morefield.

9. On the projectification of global health, see Susan Reynolds Whyte, Michael A. Whyte, Lotte Meinert, and Jenipher Twebaze, "Therapeutic Clientship: Belonging in Uganda's Projectified Landscape of AIDS Care," in *When People Come First: Critical Studies in Global Health*, ed. João Biehl and Adriana Petryna (Princeton, NJ: Princeton University Press, 2013), 140–165. On the proliferation of metrics in global health more broadly, see Vincanne Adams, ed., *Metrics: What Counts in Global Health* (Durham, NC: Duke University Press, 2016).

10. For more on the *Whole Earth Catalog*, see chapter 2.

11. Paul Ehrlich, *The Population Bomb* (New York: Ballantine Books, 1968).

12. See, e.g., Laura Briggs, *Reproducing Empire: Race, Sex, Science, and US Imperialism in Puerto Rico* (Berkeley: University of California Press, 2003).

13. For more on the history of family planning programs, see Matthew Connelly, *Fatal Misconception: The Struggle to Control World Population* (Cambridge, MA: Harvard University Press, 2008).

14. See, e.g., John Bongaarts, "A Framework for Analyzing the Proximate Determinants of Fertility," *Population and Development Review* 4, no. 1 (March 1978): 105–132. For a broad overview of these links as they pertain to multilateral policy, see UN Department of Economic and Social Affairs, Population Division, "Linkages between Population and Education: A Technical Support Services Report," December 1997, http://www.un.org/esa/population/pubsarchive/tsspop/tss976/gbc976.htm#bib.

15. See Arvonne Skelton Fraser, *The UN Decade for Women: Documents and Dialogue* (Boulder, CO: Westview Press, 1987).

16. This is reminiscent of Ruth Schwartz Cowan's work *More Work for Mother: The Ironies of Household Technology from the Open Hearth to the Microwave* (New York: Basic Books, 1983), which shows that the addition of supposedly labor-saving devices to the home did not reduce a woman's workload—rather, it increased it, as expectations for women's work rose in step. In this case, the technologies are designed to be low-tech and labor-intensive, for example, by relying on hand cranks rather than electric motors, which presumably would have a similar, if not amplified, effect on women's workloads.

17. UN Conference on Women, "Forward Looking Strategies for the Advancement of Women to the Year 2000," Nairobi, July 15–26, 1985, reprinted in the Report of Congressional Staff Advisors to the Nairobi Conference to the Committee on Foreign Affairs, para. 200, p. 127, US House of Representatives, US Government Printing Office, January 1986.

18. UN Conference on Women, para. 205, p. 128.

19. UN Conference on Women, para. 156, p. 117.

20. Connelly, *Fatal Misconception*. Eventually, this would lead to the introduction, by the Reagan administration in 1984, of the "global gag rule," which cut US government funding to any NGOs that counseled on, performed, or advocated for abortion services—even if those activities were privately funded.

21. Gordon Duncan interview with Jolayne Houtz, PATH, November 2015, PATH private internal archives.

22. "The Birth of PATH," PATH, http://www.path.org/about/birth-of-path.php.

23. See Wendy Kline, *Bodies of Knowledge: Sexuality, Reproduction, and Women's Health in the Second Wave* (Chicago: University of Chicago Press, 2010), 99, 118–199; Cass Peterson, "Indians Given Questionable Drug," *Washington Post*, August 7, 1987, A20.

24. Duncan interview with Houtz. See also "Honoring Decades of Compassion and Friendships," PATH, October 17, 2016, https://www.path.org/articles/honoring-decades-of-compassion-and-friendships/. The article includes many stories from friends and former employees, and was curated to celebrate Gordon Duncan at his death.

25. On obstetric and gynecological experiments on enslaved black women, see Deirdre Cooper Owens, *Medical Bondage: Race, Gender, and the Origins of American Gynecology* (Athens: University of Georgia Press, 2017).

26. See, e.g., PATH's Plain Talk program funded by the Ford Foundation, as described in "Final Report to the Ford Foundation: Plain Talk: Developing Health Information Materials for Low-Literate English- and Non-English Speaking Young People in the United States," Grant 880-1064, Ford Foundation Archives, Rockefeller Archive Center, RACcess #37609; Mary Beth Moore (PATH vice president) to Dr. Jose Barzellato (Ford Foundation Population Office), May 18, 1990, p. 5, Grant 880-1064, Ford Foundation Archives, Rockefeller Archive Center, RACcess #37609.

27. Heidi Morefield interview with Richard Mahoney, February 14, 2018. Interestingly, there were (non-PATH related) efforts to make contraceptives out of local yams, which employed large numbers of campesinos in Mexico in the mid-twentieth century. See Gabriella Soto-Laveaga, *Jungle Laboratories: Mexican Peasants, National Projects, and the Making of the Pill* (Durham, NC: Duke University Press, 2009).

28. Peter Redfield, "Bioexpectations: Life Technologies as Humanitarian Goods," *Public Culture* 24 (2012): 157–184.

29. Peggy Morrow interview with Jolayne Houtz, PATH, November 2015, PATH private internal archives.

30. Morefield interview with Mahoney.

31. Morefield interview with Mahoney.

32. Heidi Morefield interview with Gordon Perkin, September 28, 2016.

33. Tom Paulson, "At 5 Billion, PATH's Life-Saving Labels Make Vaccines More Effective," *Humanosphere*, July 14, 2014, http://www.humanosphere.org/science/2014/07/at-5-billion-paths-life-saving-labels-make-vaccines-more-effective/.

34. Morefield interview with Perkin.

35. Morefield interview with Mahoney.

36. Morefield, "More with Less."

37. Morefield interview with Mahoney.

38. Morefield interview with Mahoney.

39. For a critical take on this history, see Allan Brandt, *No Magic Bullet: A Social History of Venereal Disease in the United States since 1880* (New York: Oxford University Press, 1985). A popular film, titled *Dr. Ehrlich's Magic Bullet*, released in 1940, also idealized this story.

40. Morefield interview with Mahoney.

41. Morefield interview with Mahoney.

42. Morefield interview with Mahoney.

43. NGO Forum: Science and Technology for Development (pamphlet and program), August 19–29, 1979, Series X: Miscellaneous UNCSTD and UNESCO Materials, box 14, folder: NGO Forum American Association for the Advancement of Science Archives: Science, Technology, and Development Records, Washington, DC.

44. "Support Grows for Rival Conference on Development," *Nature* 279 (June 7, 1979): 465.

45. "Support Grows."

46. "Support Grows."

47. Dr. A. Karim Ahmed, Chairman, to the American Association for the Advancement of Science, "Non-Governmental Organization Committee Preparing Two-Day Forum on Technological Alternatives for Third World Countries" (memo), January 17, 1979, Series X: Miscellaneous UNCSTD and UNESCO Materials, box 14, folder: NGO Forum, American Association for the Advancement of Science Archives: Science, Technology, and Development Records, Washington, DC.

48. Ahmed to American Association for the Advancement of Science, 2.

49. I had the pleasure of visiting and staying at the Farm on two separate occasions in 2011 and 2012, and completed midwifery and neonatal resuscitation training with Ina May Gaskin. Over dinner one night, Stephen Gaskin recounted the community's origins. He passed away in 2014. Although my work has moved in a different direction, historian Wendy Kline has recently completed a book which heavily features Ina May Gaskin and the Farm; see Wendy Kline, *Coming Home: How Midwives Changed Birth* (Oxford: Oxford University Press, 2018). For more on Stephen Gaskin, see Douglas Martin, "Stephen Gaskin, Hippie Who Founded an Enduring Commune, Dies at 79," *New York Times*, July 2, 2014.

50. "About Us," Centre of Science for Villages, http://www.csvtech.in.

51. For more on VITA, the ITDG, and the *Whole Earth Catalog*, see chapter 3. See also George McRobie, *Small Is Possible* (London: Jonathan Cape, 1981).

52. Richard M. Harley, "Global Networks, Trading Recipes and Technologies from Maine to Nepal," *Christian Science Monitor*, October 7, 1982, https://www.csmonitor.com/1982/1007/100729.html.

53. Dr. William Ellis, "Flapping Butterfly Wings: A Retrospective of TRANET's First Twenty Years," ed. Hildegarde Hannum (Eighteenth Annual E. F. Schumacher Lecture, October 1998, Salisbury, CT).

54. Ellis.

55. Ellis, "Flapping Butterfly Wings."

56. Ellis.

57. Harley, "Global Networks."

58. Harley.

59. For more on VITA and the ITDG, see chapter 2.

60. The standard reference for the history of the internet remains Janet Abbate's *Inventing the Internet* (Cambridge, MA: MIT Press, 1999). Recently, there has been a push to disentangle histories of computer networking from histories of the internet specifically—the former being a much broader and less linear field. See especially Andrew Russell's working paper from the 2012 meeting of the Special Interest Group on Computers, Information, and Society (SIGCIS) of the Society for the History of Technology, "Histories of Networking vs. the History of the Internet," http://arussell.org/papers/russell-SIGCIS-2012.pdf. He advances similar lines of argument in Andrew Russell and Valérie Schafer, "In the Shadow of ARPANET and Internet: Louis Pouzin and the Cyclades Network in the 1970s," *Technology and Culture* 55 (2014): 880–907, as well as in his book, Andrew Russell, *Open Standards and the Digital Age: History, Ideology, and Networks* (Cambridge: Cambridge University Press, 2014). On the history of computer networking as it relates to the history of Silicon Valley, see O'Mara, *The Code* (2019).

61. Ellis, "Flapping Butterfly Wings."

62. Ellis.

63. Oscar Harkavy to John Foster-Bey, January 12, 1988, microfilm reel L-536, grant no. L-87–214, Program for Appropriate Technology in Health, Ford Foundation Grant Files, Rockefeller Archive Center.

64. Harkavy to Foster-Bey.

65. See Dr. Julia Walsh and Dr. Kenneth Warren, "Selective Primary Health Care: An Interim Strategy for Disease Control in Developing Countries," *New England*

Journal of Medicine 308, no. 18 (November 1, 1979): 967–974. This paper was originally presented at the Bellagio, Italy, conference in April 1979.

66. Advanced through the WHO's 1977 Model List of Essential Drugs and published every two years since, essential drugs (today, essential medicines) are those that are "basic, indispensable, and necessary to the health of the population." See WHO, *The Selection of Essential Drugs* (WHO Technical Reports Series No. 615), Geneva, 1977, 9. For a detailed history of the essential drugs concept in global health, see Jeremy Greene, "Making Medicines Essential: The Emergent Centrality of Pharmaceuticals in Global Health," *BioSocieties* 6, no. 1 (2011): 10–33. The projects in question included Project for Strengthening Health Delivery Systems in Central and West Africa (1979), the sectoral survey Pharmaceuticals and Health Care Products in Egypt (1982), Preparation of a Primary Health-Care Formulary on guidance documents for USAID missions (1983), and the Latin American and Caribbean regional project Technology Development and Transfer in Health, which had a sizable essential drugs component and regional strategy (1984). The 1985 Program Guidance Paper on Pharmaceuticals in Health Assistance identified seventy-three different USAID health projects with pharmaceutical components in 1983 alone. All project reports can be found in the Development Experience Clearinghouse, USAID.

67. Greene, "Making Medicines Essential"; Fiona Godlee, "WHO in Retreat: Is It Losing Its Influence?" *British Medical Journal* 309 (1994): 1491–1495; Gill Walt and Jan Willem Harnmeijer, "Formulating an Essential Drugs Policy: WHO's Role," in *Drugs Policy in Developing Countries*, by Najmi Kanji, Anita Hardon, Jan Willem Harnmeijer, Masuma Mamdani, and Gill Walt (London: Zed Books, 1992).

68. Marcos Cueto, "The Origins of Primary Health Care and Selective Primary Health Care," *American Journal of Public Health* 94 (2004): 1864–1874.

69. USAID, "Technology for Primary Health Care: Project Data Sheet, Amendment 1" (1985), Development Experience Clearinghouse, USAID.

70. Heidi Morefield, oral history interview with Buck Greenough, Baltimore, May 16, 2018.

71. Management Sciences for Health, "The PRITECH Project: Second Annual Report, Oct. 1, 1984–Sept. 30, 1985," Development Experience Clearinghouse, USAID.

72. Management Sciences for Health, "PRITECH II: Technical Advisory Meeting," February 16, 1990, 77, 84, Development Experience Clearinghouse, USAID.

73. Management Sciences for Health, 19, 21, 22, 29.

74. Management Sciences for Health, 7–8.

75. PATH, "HealthTech: Technologies for Child Health" (brochure), 1989, folder A94 PS9022, 1989–1992, Program for Appropriate Technology in Health, Ford 76. Morefield interview with former PATH project manager who wished to remain anonymous, July 5, 2018. See also the many works of Paul Farmer, especially *Infections and Inequalities: The Modern Plagues* (Berkeley: University of California Press, 1999).

77. William Muraskin, *The War against Hepatitis B: A History of the International Task Force on Hepatitis B Immunization* (Philadelphia: University of Pennsylvania Press, 1995).

78. See "Our Founders," Partners in Health, https://www.pih.org/our-founders. The documentary film *Bending the Arc* (2017), directed by Kief Davidson and Pedro Kos, also tells the story of PIH's founding and philosophy.

79. Paul Farmer passed away in February 2022. For more on his life and work, see Tracy Kidder, *Mountains beyond Mountains: The Quest of Dr. Paul Farmer, a Man Who Would Cure the World* (New York: Random House, 2003).

CHAPTER SEVEN

1. ZANU-PF, "Appropriate Technology Demonstration Village—Melfort," *Zimbabwe News*, July 1, 1989, cover.

2. For more on Etawah and model villages, see chapter 1.

3. After the party's Women's League decided it would put forth a female candidate for the vice presidency and the motion was supported by a party faction led by Mujuru's husband, General Solomon Mujuru, Joice Mujuru was sworn in as Zimbabwe's vice president on December 6, 2004. She held the position until she was expelled from the party in December 2014 for allegedly plotting against Mugabe—a charge leveled primarily by the president's wife, Grace Mugabe, who wanted to position herself to succeed her ailing husband. After Mugabe was himself expelled from the presidency in 2017 by ZANU-PF officials loyal to Emerson Mnangagwa, longtime speaker of Parliament and Mujuru's successor as vice president, who did not want Grace Mugabe as party leader, Mujuru received a formal apology from Robert Mugabe, and the two ex-comrades remain on good terms. She unsuccessfully challenged Mnangagwa for the presidency in the July 2018 election. See Blessing Zulu, "Ousted Mugabe and Joice Mujuru Reconnect, Forgive, Laugh over Deserting Friends," *VOA Zimbabwe*, February 2, 2018, https://www.voazimbabwe.com/a/robert-mugabe-joice-mujuru-reconnect-forgive-and-laugh/4235102.html.

4. Shona is a Bantu language spoken by about 70 percent of the population in Zimbabwe.

5. ZANU-PF, "New Home-Building Techniques Set to Transform Rural Life," *Zimbabwe News*, July 1, 1989, 14.

6. ZANU-PF, "Mujuru Call for Appropriate Technology to Lessen Women's Heavy Load," *Zimbabwe News*, July 1, 1989, 12.

7. ZANU-PF, "New Home-Building Techniques," 13.

8. UN Conference on Women, "Forward Looking Strategies for the Advancement of Women to the Year 2000," Nairobi, July 15–26, 1985, reprinted in the Report of Congressional Staff Advisors to the Nairobi Conference to the Committee on Foreign Affairs, para. 259, 138–140, US House of Representatives. US Government Printing Office, January 1986. A footnote notes that the United States voted against the inclusion of this paragraph in the Forward Looking Strategies, as it opposed the imposition of sanctions and aid to liberation movements.

9. Martha F. Loutfi, *Rural Women: Unequal Partners in Development* (Geneva: International Labour Office, 1980), 1, 21.

10. For more on the NIEO negotiations, see chapter 3.

11. For more on the ITDG's field office operations, see chapter 4.

12. David Engerman, "Development Politics and the Cold War," *Diplomatic History* 41, no. 1 (2017): 2.

13. James Ferguson, *The Anti-Politics Machine: Development, Depoliticization, and Bureaucratic Power in Lesotho* (Minneapolis: University of Minnesota Press, 1994).

14. For more on CASTAFRICA, see chapter 3.

15. Peter Morgan, "Recent Developments in Environmental Sanitation and their

Role in the Prevention of Bilharziasis," *Central African Journal of Medicine* 23, no. 11 (1977): 11–15, at 15.

16. Suzanne de Laet and Annemarie Mol, "The Zimbabwe Bush Pump: Mechanics of a Fluid Technology," *Social Studies of Science* 30, no. 2 (April 2000): 225–263, at 251–252.

17. Clapperton Mavhunga, *Transient Workspaces: Technologies of Everyday Innovation in Zimbabwe* (Cambridge, MA: MIT Press, 2014).

18. R. T. Mossop, "College of General Practice 1980 Robbie Gibson Memorial Lecture: Doctors—Health Promoters or Sickness Professionals?" *Central African Journal of Medicine* 26, no. 11 (November 1980): 231–235.

19. See UN General Assembly Resolution, "Question of Southern Rhodesia," Resolutions 2022 (November 5, 1965), 2105 (December 20, 1965), 2138 (October 22, 1966), and 2151 (November 17, 1966).

20. For more on ways that US support of Rhodesia continued in spite of public-facing sanctions, see Vaughan E. Taplin, "US Support of Rhodesia," *Black Scholar* (February 1971): 51–55; Eddie Michel, "'This Outcome Gives Me No Pleasure. It Is Extremely Painful for Me to Be the Instrument of Their Fate': White House Policy on Rhodesia during the UDI Era (1965–1979)," *South African Historical Journal* 71, no. 3 (2018): 442–465.

21. Charles Mohr, "Whites' Fear of Black Rebels Spreading in Rhodesia," *New York Times* May 13, 1973, 1.

22. Allan Savory, "Ramblings," September 15, 2001, http://theconversation.org/archive/ramblings.html.

23. Mugabe called the official political wing of ZANU the ZANU-PF (Zimbabwe African National Union—Patriotic Front).

24. R. S. Summers, "How Essential Is the WHO 'Essential' Drug List? The Example of Zimbabwe," *Central African Journal of Medicine* 27, no. 11 (1981): 228–231. See also the follow-up article: R. S. Summers, "PEDLIZ: Unwarranted Restriction of Prescribing Freedom or Rational Cost-Effective Measure?" *Central African Journal of Medicine* 29, no. 2 (February 1983): 43–46. "PEDLIZ" refers to the Proposed Essential Drug List for Zimbabwe.

25. Theories of an "epidemiological transition," wherein developing countries are thought to be still grappling with infectious diseases while developed countries primarily have to deal with chronic diseases (sometimes characterized as diseases of "civilization" or of "modernity") have led to underinvestment in areas of medicine such as oncology in the Global South. This is slowly starting to change as more attention is given to noninfectious diseases in these regions. See, for example, Julie Livingston, *Improvising Medicine: An African Oncology Ward in an Emerging Cancer Epidemic* (Durham, NC: Duke University Press, 2012).

26. "Self-Sufficiency as a Rallying Point for Economic Independence," *Zimbabwe News*, January 1987.

27. UNESCO, *Second Conference of Ministers Responsible for the Application of Science and Technology to Development in Africa, CASTAFRICA II: Final Report* (Arusha, Tanzania, July 6–15, 1987), 37.

28. Esther Boserup, *Woman's Role in Economic Development* (London: Compton Printing, 1970).

29. Boserup, 160–162. On the politics and process of "counting" women's labor, see also Luke Messac, "Outside the Economy: Women's Work and Feminist Economics in

the Construction of National Income Accounting," *Journal of Imperial and Commonwealth History* 46, no. 3 (2018): 552–578.

30. *Zimbabwe News*, July 1989, 12.

31. Goran Hyden (Nairobi) to William Carmichael and Peter Geithner (New York), "Proposal to Establish a Sub-Office in Zimbabwe" (memo), June 16, 1982, box 1, folder 34, Developing Country Program, Vice President, Office Files of William Carmichael (FA711), Ford Foundation Records, Rockefeller Archive Center.

32. Hyden to Carmichael and Geithner.

33. For more on the ITDG, see chapter 2.

34. "Tinker, Tiller, Technical Change," July 12, 1988, microfilm reel 5751, grant no. 885-0823, Intermediate Technology Development Group, Ford Foundation Grant Files, Rockefeller Archive Center. The title, presumably, is a play on John le Carré's 1974 novel *Tinker, Tailor, Soldier, Spy*. Although appropriate technology programs were no longer in vogue in the mid-1980s in the United States and United Kingdom, they continued to receive funding through foundations like Ford and even USAID, especially when tied to small-scale industry and "self-sufficiency."

35. "Tinker, Tiller, Technical Change," 2.

36. Marilynn Carr, "Technologies for Rural Women: Impact and Dissemination," in *Technology and Rural Women: Conceptual and Empirical Issues*, ed. Iftikhar Ahmed (London: George Allen & Unwin, 1985), 116.

37. Ministry of Community and Co-operative Development, *Women's Social and Cultural Situation*, vol. 4 of *Building Whole Communities* (Harare: Ministry of Community and Co-operative Development, 1992), 43.

38. Ministry of Community and Co-operative Development.

39. Ministry of Community and Co-operative Development, *Organising for the Future*, vol. 7 of *Building Whole Communities* (Harare: Ministry of Community and Co-operative Development, 1992), 47.

40. Ministry of Community and Co-operative Development, 55.

41. Blessing-Miles Tendi, "Robert Mugabe and Toxicity: History and Context Matter," *Representation: Journal of Representative Democracy* 47 (2011): 307–318.

42. Pedzisai Maedza, "'Gukurahundi—A Moment of Madeness': Memory Rhetorics and Remembering in the Postcolony," *African Identities* 17 (2019): 175–190.

43. Elizabeth Hull, "The Renewal of Community Health under the KwaZulu 'Homeland' government," *South African Historical Journal* 64, no. 1 (2012): 22–40. For more on this reconceptualization of the history of homeland rule, see William Beinart, "Beyond 'Homelands': Some Ideas about the History of African Rural Areas in South Africa," *South African Historical Journal* 64, no. 1 (2012): 5–21.

44. Howard Phillips, "The Return of the Pholela Experiment: Medical History and Primary Health Care in Post-Apartheid South Africa," *American Journal of Public Health* 104, no. 10 (2014): 1872–1876. See also Theodore M. Brown and Elizabeth Fee, "Sidney Kark and John Cassel: Social Medicine Pioneers and South African Émigrés," *American Journal of Public Health* 92, no. 11 (2002): 1744–1745.

45. S. M. Tollman, "Pholela Health Centre—The Origins of Community-Oriented Primary Health Care (COPC), an Appreciation of the Work of Sidney and Emily Kark," *South African Medical Journal* 84, no. 10 (1994): 653–658.

46. Abigail H. Neely, "Reconfiguring Pholela: Understanding the Relationships between Health and Environment from the 1930s to the 1980s" (PhD diss., University of Wisconsin–Madison, 2011).

47. M. Susser, "Pioneering Community-Oriented Primary Care," *Bulletin of the World Health Organization* 77, no. 5 (1999): 436–438.

48. Melissa Diane Armstrong, "The ANC's Medical Trial Run: The Anti-Apartheid Medical Service in Exile, 1964–1990" (PhD diss., Carleton University, Ottawa, 2017).

49. Melissa Diane Armstrong, *An Ambulance on Safari: The ANC and the Making of a Health Department in Exile* (Montreal, QC: McGill-Queen's University Press, 2020).

50. Timothy Gibbs, "Chris Hani's 'Country Bumpkins': Regional Networks in the African National Congress Underground, 1974–1994," *Journal of Southern African Studies* 37, no. 4 (2011): 677–691.

51. Tom Lodge, *Mandela: A Critical Life* (Oxford: Oxford University Press, 2006), 29.

52. Timothy Gibbs, *Mandela's Kinsmen: Nationalist Elites and Apartheid's First Bantustan* (Rochester, NY: Boydell and Brewer, 2014).

53. Cecil Cook, "Making and Unmaking Poverty in South Africa," *South African New Economics Network: SANE Views* 2, no. 10 (April 2002).

54. This is a similar goal as the earlier *swadeshi* technology movement in India, described by David Arnold, *Everyday Technology: Machines and the Making of India's Modernity* (Chicago: University of Chicago Press, 2013), and David Engerman, "Development Politics and the Cold War," *Diplomatic History* 41, no. 1 (2017).

55. Cecil Cook, "Pie in the Transkei: (The Search for) or (Obstacles to) an Appropriate Technology Built Environment in the Transkei," *Architecture South Africa* 34, no. 9 (1988): 64, 80.

56. Cook.

57. Heidi Morefield telephone interview with Richard Rottenburg, April 27, 2018.

58. Heidi Morefield telephone interview with Lesley Steele, June 4, 2018.

59. Morefield interview with Steele.

60. Morefield interview with Rottenburg.

61. Morefield interview with Rottenburg.

62. Cook did not carry out orders on the apartheid government's behalf, but he did conform to their demands in what he did and did not pursue. "Just following orders" has famously been characterized by Hannah Arendt as part of the "banality of evil," in that everyday, legally sanctioned orders committed by regular people in the service of institutionalized racism and white supremacy is what enables those regimes and systems to function. See Hannah Arendt, *Eichmann in Jerusalem: A Report on the Banality of Evil* (New York: Viking Press, 1963).

63. Rottenburg recalled the year this happened as 1986; however, official reports have placed the attack in June 1985. See Archbishop Desmond Tutu, Chairperson, "Appendix: National Chronology," in *Truth and Reconciliation Commission of South Africa Report, Volume 3*, October 29, 1998, 23, https://www.sahistory.org.za/sites/default/files/trc_report_volume_3.pdf; Command of Umkhonto we Sizwe, "MK in Combat," n.d., 18, http://disa.ukzn.ac.za/sites/default/files/pdf_files/Dav10n186.1681.5785.010.001.1986.5.pdf.

64. Morefield interview with Rottenburg.

65. Morefield interview with Rottenburg.

66. Morefield interview with Steele.

67. "Agencies, Activities, and Research in and on the Transkei," *Development Southern Africa* 1, no. 2 (1984): 239–245.

68. Wilma Stassen, "Old Transkei Is International Cancer 'Hotspot,'" *Health-E*

News South Africa, February 26, 2014, https://www.health-e.org.za/2014/02/26/old-transkei-international-cancer-hot-spot/?utm_source=rss&utm_medium=rss&utm_campaign=old-transkei-international-cancer-hot-spot.

69. Reports noted that when the grinding stones were sufficiently worn down, they became concave, allowing silica dust to pool in high quantities with the maize. See Vikash Sewram, "Silica Content of Homegrown Maize Linked to Cancer Risk" (paper presented at AORTIC International Cancer Conference, Cairo, November 30–December 2, 2011), https://ecancer.org/conference/137-aortic-2011/video/1246/silica-content-of-home-ground-maize-linked-to-cancer-risk.php.

70. Khathatso Mokoetle, personal correspondence with author, August 31, 2018.

71. Khathatso Mokoetle and Barbara Klugman, "Remobilising Civil Society for Sexual Rights: The Establishment of SHARISA," *Agenda: Empowering Women for Gender Equity* 26, no. 2 (2012): 15–23.

72. Heidi Morefield interview with Peter Rampora, March 26, 2018, Johannesburg, South Africa.

73. NPPHCN, "Our Mission," box A, folder 1, NPPHCN Papers AG3176, University of Witwatersrand Historical Papers, Johannesburg, South Africa. See also NPPHCN, "National Progressive Primary Health Care Network" (brochure), 6, box A, folder 1, NPPHCN Papers AG3176. On the transition from primary health care to selective primary health care, see Marcos Cueto, "The Origins of Primary Health Care and Selective Primary Health Care," *American Journal of Public Health* 94, no. 11 (2004): 1864–1874.

74. NPPHCN, "National Progressive Primary Health Care Network, PHC Activity Areas chart," box A, folder 1, NPPHCN Papers AG3176.

75. NPPHCN, "National Progressive Primary Health Care Network."

76. NPPHCN, "National Progressive Primary Health Care Network," 4.

77. NPPHCN, "National Progressive Primary Health Care Network," 9.

78. NPPHCN, "Health for the Rainbow Nation: The Story of the National Progressive Primary Health Care Network," 3, box A, folder 1, NPPHCN Papers AG3176.

79. NPPHCN, "Health for the Rainbow Nation," 7–8.

80. NPPHCN, "Health for the Rainbow Nation," 9.

81. NPPHCN, "Health for the Rainbow Nation," 16.

82. Howard Phillips, "The Return of the Pholela Experiment: Medical History and Primary Health Care in Post-Apartheid South Africa," *American Journal of Public Health* 104, no. 10 (2014): 1872–1876.

83. Heidi Morefield telephone interview with Dr. Irwin Friedman, August 22, 2018. Nkosazana Dlamini-Zuma is still in government and currently holds the post of minister in the presidency for the National Planning Commission for Policy and Evaluation.

84. NPPHCN, *Community Radio*, March 1997, box G, folder 12, NPPHCN Papers AG3176.

85. NPPHCN, *Community Radio*, October 1998, box G, folder 12, NPPHCN Papers AG3176.

CHAPTER EIGHT

1. Jane Jacobs, "Downtown Is for People," *Fortune*, 1958, http://fortune.com/2011/09/18/downtown-is-for-people-fortune-classic-1958/.

2. On productivity in modern culture, see Melissa Gregg, "The Productivity

Obsession," *The Atlantic*, November 13, 2015, https://www.theatlantic.com/business/archive/2015/11/be-more-productive/415821/.

3. See, e.g., Warwick Anderson, "Excremental Colonialism: Public Health and the Poetics of Pollution," *Critical Inquiry* 21, no. 3 (Spring 1995): 640–669; Yu Xinzhong, "The Treatment of Night Soil and Waste in Modern China," in *Health and Hygiene in Chinese East Asia: Policies and Public in the Long Twentieth Century*, ed. Charlotte Furth and Angela Ki Che Leung (Chapel Hill, NC: Duke University Press, 2010), 51–72.

4. A clip of the episode, from January 22, 2015, is available at "Bill Gates and Jimmy Drink Poop Water," YouTube video, 3:56, posted by Tonight Show Starring Jimmy Fallon, https://www.youtube.com/watch?v=FHgsL0dpQ-U.

5. See, e.g., David Goldman, "Bill Gates' Poop Water Machine Gets a Test Run," *CNN Money*, August 13, 2015, https://money.cnn.com/2015/08/13/technology/bill-gates-poop-water/index.html.

6. Carol Tice, "To Stay Lean, Gates Foundation Outsources," *Puget Sound Business Journal*, July 4, 1999, https://www.bizjournals.com/seattle/stories/1999/07/05/newscolumn1.html. The article opens "At first glance, the relationship between the William H. Gates Foundation and the Program for Appropriate Technology in Health seems almost suspect." It goes on to call the relationship "incestuous."

7. As quoted in Anand Giridharadas, *Winners Take All: The Elite Charade of Changing the World* (New York: Vintage Books, 2018), 73.

8. See "About Us," Masters of Scale, https://mastersofscale.com/about-us/. On where and how Bill Gates does his "big thinking," see Reid Hoffman, "How to Find Your Big Idea," Masters of Scale, https://mastersofscale.com/sara-blakely-how-to-find-your-big-idea/. Gates is also famously a fan of "big history"; see Bill Gates, "Lifelong Learners Will Appreciate This Book about the History of Everything," *GatesNotes* (blog), May 21, 2018, https://www.gatesnotes.com/Books/Origin-Story?WT.mc_id=20180529134311_SummerBooks2018OriginStory_BG-FB&WT.tsrc=BGFB&linkId=52301001.

9. Tom Paulson, "Suzanne Cluett: 1942–2006: Her Mission Started with the Peace Corps," *Seattle Post-Intelligencer*, January 8, 2006.

10. Heidi Morefield interview with Richard Mahoney, February 14, 2018.

11. Paulson, "Suzanne Cluett."

12. Paulson.

13. The dinner apparently came to be known colloquially as "the lamb chop dinner" among many attendees, due to the memorable entrée. See Michael Barbaro, "Can Bill Gates Vaccinate the World?" *The Daily* (podcast), March 3, 2021, transcript at https://www.nytimes.com/2021/03/03/podcasts/the-daily/coroanvirus-vaccine-bill-gates-covax.html?showTranscript=1.

14. Morefield interview with Mahoney.

15. As quoted in Barbaro, "Can Bill Gates Vaccinate the World?"

16. Giridharadas, *Winners Take All*, 74.

17. Dan Goodin, "Revisiting the Spectacular Failure That Was the Bill Gates Deposition," *Ars Technica*, September 10, 2020, https://arstechnica.com/tech-policy/2020/09/revisiting-the-spectacular-failure-that-was-the-bill-gates-deposition/.

18. See Barbaro, "Can Bill Gates Vaccinate the World?"

19. TRIPS came into effect January 1, 1995, and grew out of the General Agreement on Tariffs and Trade (GATT). Historian Quinn Slobodian has argued that GATT is the culmination of "Geneva School" neoliberals' efforts to create a world institution to gov-

ern the global economy and, in the interim, curtail the demands of the world's poorer countries channeled through the NIEO. See Quinn Slobodian, "A World of Signals," in *Globalists: The End of Empire and the Birth of Neoliberalism* (Cambridge, MA: Harvard University Press, 2018), ch. 7.

20. The original challenges were to (1) create effective single-dose vaccines for use soon after birth; (2) prepare vaccines that did not require refrigeration; (3) develop needle-free delivery systems for vaccines; (4) devise reliable tests in model systems to evaluate live-attenuated vaccines; (5) solve the design of antigens for effective, protective immunity; (6) learn which immunological responses provide protective immunity; (7) develop a genetic strategy to deplete or incapacitate a disease-transmitting insect population; (8) develop a chemical strategy to deplete or incapacitate a disease-transmitting insect population; (9) create a full range of optimal, bioavailable nutrients in a single staple plant species; (10) discover drugs and delivery systems that minimize the likelihood of drug resistant microorganisms; (11) create therapies that cure latent infection; (12) create immunological methods that cure latent infection; (13) develop technologies that permit quantitative assessment of population health; and (14) develop technologies that allow assessment of individuals for multiple conditions or pathogens at point of care. See "Read the Grand Challenges," Grand Challenges in Global Health, 2011, https://web.archive.org/web/20111224114326/http://www.grandchallenges.org/Pages/BrowseByGoal.aspx.

21. For an example of the criticism, see Anne-Emanuelle Birn, "Gates's Grandest Challenge: Transcending Technology as Public Health Ideology," *The Lancet* 366 (2005): 514–519.

22. For comparison, the United States typically contributes around 20 percent of the WHO's budget, although in 2019 this was reduced to 15 percent because of the Trump administration's withdrawal. The Gates Foundation's contribution is the second largest after the United States. WHO budget data is available at http://open.who.int/2018-19/contributors/contributor. Gates Foundation budgets are listed at https://www.gatesfoundation.org/Who-We-Are/General-Information/Foundation-Factsheet. For more on how the Gates Foundation vets and steers the agenda of even the World Health Organization, see Linsey McGoey, *No Such Thing as a Free Gift: The Gates Foundation and the Price of Philanthropy* (Brooklyn, NY: Verso Books, 2015).

23. Heide Morefield interview with Gordon Perkin, September 28, 2016.

24. Geoff Watts, "Gordon Wesley Perkin" (obituary), *The Lancet* 397 (January 9, 2021): 91.

25. Heidi Morefield interview with Chris Elias, Gates Foundation, February 8, 2018.

26. Janet Echelman (Boston), *Impatient Optimist*, spliced and braided fiber with colored lighting, Gates Foundation Campus, Seattle.

27. Morefield interview with Elias.

28. Morefield interview with Elias.

29. Morefield interview with Elias.

30. Morefield interview with Elias.

31. On the name PATH, Chris Elias told me: "Gordon [Perkin] is famous for his acronyms. He delights in forming acronyms. I know he does. He thinks about it a lot . . . the Global Alliance for Improved Nutrition, 'GAIN.' That was his." Morefield interview with Elias.

32. On why innovative coolers will not solve the ongoing Ebola outbreak in Democratic Republic of the Congo, see Heidi Morefield, "If We Really Want to Eradicate

Diseases Like Ebola, We Need a New Strategy," *Washington Post*, July 9, 2018, https://www.washingtonpost.com/news/made-by-history/wp/2018/07/09/if-we-really-want-to-eradicate-diseases-like-ebola-we-need-a-new-strategy/?utm_term=.80ea436f61f5.

33. Jeff Bernson, Darin Zehrung, Bhavya Gowda, and Michelle Mulder, "Five Years Later: The Global Health Innovation Accelerator," PATH, December 12, 2019, https://www.path.org/articles/five-years-later-global-health-innovation-accelerator/.

34. "How Ellavi Works," Ellavi, http://ellavi.com/#section-how-ellavi-works.

35. Tom Vanderbilt, "'Reverse Innovation' Could Save Lives. Why Aren't We Using It?" *New Yorker*, February 4, 2019.

36. "The Need by the Numbers," Ellavi, http://ellavi.com/#section-how-ellavi-works.

37. Petra Brhlikova, Patricia Jeffery, G. P. Bhatia, and Sakshi Khurana, "Intrapartum Oxytocin (Mis)use in South Asia," *Journal of Health Studies* (2009): 33–50, http://r4d.dfid.gov.uk/PDF/Outputs/ESRC_DFID/60336_Intrapartum.pdf.

38. PATH, "Lifesaving Medical Device for Postpartum Hemorrhage Receives Regulatory Approval in Ghana and Kenya" (press release), July 28, 2020, https://www.path.org/media-center/lifesaving-medical-device-postpartum-hemorrhage-receives-regulatory-approval-ghana-and-kenya/.

39. Elizabeth Abu-Haydar to Heidi Morefield, March 19, 2018, email correspondence following interview on February 26, 2018.

40. Heidi Morefield interview with SAMRC employee who wished to remain anonymous, March 20, 2018.

41. Heidi Morefield interview with Tony Bunn, Cape Town, March 19, 2018.

42. Morefield interview with Bunn.

43. Morefield interview with Bunn; Morefield interview with anonymous SAMRC employee.

44. Heidi Morefield, personal correspondence with Tony Bunn, March 29, 2021.

45. Morefield correspondence with Bunn.

46. Heidi Morefield interview with anonymous interviewee, Johannesburg, March 23, 2018. The interview subject did not wish for his name to be used.

47. "Technology Justice," Practical Action, https://policy.practicalaction.org/policy-themes/technology-justice.

48. On the concept of "insulating" institutions that "encase" markets in the neoliberal framework, see Slobodian, *Globalists*, 4–7.

49. Dave Lee, "Why Big Tech Pays Poor Kenyans to Teach Self-Driving Cars," *BBC News*, November 3, 2018, https://www.bbc.com/news/technology-46055595; Louise Bezuidenhout, "A New Way to Equip Africa's Science Labs: Get Students to Build Their Own," *The Conversation*, August 5, 2018, https://theconversation.com/a-new-way-to-equip-africas-science-labs-get-students-to-build-their-own-100792.

50. Johns Hopkins University has a medical-student-run organization, Supporting Hospitals Abroad with Resources and Equipment (SHARE), that recovers "clean, unused medical supplies" at Johns Hopkins Hospital that would otherwise be discarded and donates them to "resource-poor hospitals" in the Global South. See Catherine Arnold, "In Short, Supplies," *Johns Hopkins Magazine*, Winter 2014, https://hub.jhu.edu/magazine/2014/winter/medical-supplies-waste/.

51. Theodore Brown, Marcos Cueto, and Elizabeth Fee, "The World Health Organization and the Transition from 'International' to 'Global' Public Health," *American Journal of Public Health* 96 (2006): 62–72.

52. Allan Brandt, "How AIDS Invented Global Health," *New England Journal of Medicine* 368 (2013): 2149–2152.

53. Dr. Halfdan Mahler, "Address to the 61st World Health Assembly," Geneva, May 19, 2008, http://www.who.int/mediacentre/events/2008/wha61/hafdan_mahler_speech/en/.

EPILOGUE

1. Evgeny Morozov, "The Tech 'Solutions' for Coronavirus Take the Surveillance State to the Next Level," *The Guardian*, April 15, 2020.

2. For more on the literature of health disparities in the United States, see, e.g., Keith Wailoo, *Dying in the City of the Blues: Sickle Cell Anemia and the Politics of Race and Health* (Chapel Hill: University of North Carolina Press, 2001); Alondra Nelson, *Body and Soul: The Black Panther Party and the Fight against Medical Discrimination* (Minneapolis: University of Minnesota Press, 2011); Jim Downs, *Sick from Freedom: African-American Illness and Suffering during the Civil War and Reconstructions* (New York: Oxford University Press, 2012); Anne Pollock, *Medicating Race: Heart Disease and Durable Preoccupations with Difference* (Durham, NC: Duke University Press, 2012).

3. Tiffany N. Ford, Sarah Reber, and Richard V. Reeves, "Race Gaps in COVID-19 Deaths Are Even Bigger Than They Appear," *Up Front* (blog), Brookings Institute, June 16, 2020, https://www.brookings.edu/blog/up-front/2020/06/16/race-gaps-in-covid-19-deaths-are-even-bigger-than-they-appear/.

4. Stephanie Soucheray, "US Blacks 3 Times More Likely Than Whites to Get COVID-19," Center for Infectious Disease Research and Policy, University of Minnesota, August 14, 2020, https://www.cidrap.umn.edu/news-perspective/2020/08/us-blacks-3-times-more-likely-whites-get-covid-19.

5. Gbenga Ogedegbe, Joseph Ravenell, and Samrachana Adhikari, "Assessment of Racial/Ethnic Disparities in Hospitalization and Mortality in Patients with COVID-19 in New York City," *JAMA Network Open*, December 4, 2020, https://pubmed.ncbi.nlm.nih.gov/33275153/.

6. Morozov, "Tech 'Solutions.'"

7. On the high-tech imperative in modern US medicine, see Deborah Levine, "No, Mr. President, Healthcare Workers Aren't Stealing Masks. You Failed Them," *Washington Post*, March 30, 2020, https://www.washingtonpost.com/outlook/2020/03/30/low-tech-medical-equipment-saves-lives/ok/2020/03/30/low-tech-medical-equipment-saves-lives/.

8. Stephen Collinson, "Trump Passes the Buck as Deadly Ventilator Shortage Looms," *CNN*, April 3, 2020, https://www.cnn.com/2020/04/02/politics/donald-trump-ventilators-leadership-coronavirus/index.html; Courtney Subramanian, "Governors Warn of Dire Ventilator Shortages as Virus Pandemic Rages. Trump Says Some Are Playing 'Politics,'" *USA Today*, April 4, 2020, https://www.usatoday.com/story/news/politics/2020/04/04/coronavirus-trump-says-states-playing-politics-ventilator-requests/5111963002/.

9. Ruth Maclean and Simon Marks, "10 African Countries Have No Ventilators. That's Only Part of the Problem," *New York Times*, April 18, 2020.

10. Farai D. Madzimbamuto, "Ventilators Are Not the Answer in Africa," *African*

Journal of Primary Health Care and Family Medicine 12 (2020): 2517; Efosa Ojomo, "African Countries Don't Need Donated Ventilators without Revamping Their Health Systems First," *Quartz Africa*, May 2, 2020, https://qz.com/africa/1850216/africa-doesnt-need-jack-mas-ventilators-building-health-systems/.

11. Leah Rosenbaum, "As Coronavirus Spreads Globally, These Researchers Are Designing Ventilators That Cost Less Than $1000," *Forbes*, April 30, 2020, https://www.forbes.com/sites/leahrosenbaum/2020/04/30/as-coronavirus-spreads-globally-these-researchers-are-designing-ventilators-that-cost-less-than-1000/?sh=6ab8555d5cb2.

12. Andrew Good, "NASA Develops COVID-19 Prototype Ventilator in 37 Days," NASA Jet Propulsion Laboratory, April 23, 2020, https://www.nasa.gov/feature/jpl/nasa-develops-covid-19-prototype-ventilator-in-37-days.

13. Shomari Stone, "Maryland Engineers Converting Breast Pumps into Ventilators," *NBC Washington*, April 7, 2020, https://www.nbcwashington.com/news/local/maryland-engineers-converting-breast-pumps-into-ventilators/2267494/; "Can Our Breast Pump Become a Ventilator?" Medela Engineering Challenge, https://www.medela.com/company/news/news/medela-engineering-challeng-breast-pump-into-ventilator.

14. See Anne Pollock, *Synthesizing Hope: Matter, Knowledge, and Place in South African Drug Discovery* (Chicago: University of Chicago Press, 2019).

15. Sarah Wild, "South Africa Is Working on Producing 10,000 Ventilators by the End of June," *Quartz Africa*, April 8, 2020, https://qz.com/africa/1835025/south-africa-producing-10000-ventilators-for-covid-19-by-june/; Wendell Roelf, "South Africa Produces Its First Ventilators to Fight COVID-19," *Reuters*, July 31, 2020, https://www.reuters.com/article/us-health-coronavirus-safrica-ventilator/south-africa-produces-its-first-ventilators-to-fight-covid-19-idUSKCN24W1VD.

16. SAMRC, "Presentation on the Launch of MeDDIC," PowerPoint, March 26, 2021, received via email correspondence with Tony Bunn.

17. See Michael Barbaro, "Can Bill Gates Vaccinate the World?" *The Daily* (podcast), March 3, 2021, transcript at https://www.nytimes.com/2021/03/03/podcasts/the-daily/coroanvirus-vaccine-bill-gates-covax.html?showTranscript=1.

18. Barbaro.

19. Barbaro.

20. Seth Berkley, "COVAX Explained," Gavi, the Vaccine Alliance, September 3, 2020, https://www.gavi.org/vaccineswork/covax-explained

21. Hannah Kettler, "What Is COVAX?" PATH, February 5, 2021, https://www.path.org/articles/what-covax/?gclid=CjwKCAjwjbCDBhAwEiwAiudByxyG-PYalh8Sa8VdSlaFcIblYLltL59DjeBZHnyBMRyF2nutJhDxqBoC4gwQAvD_BwE.

22. Barbaro, "Can Bill Gates Vaccinate the World?"

23. Jay Hancock, "They Pledged to Donate Rights to Their COVID Vaccine, Then Sold Them to Pharma," *Kaiser Health News*, August 25, 2020, https://khn.org/news/rather-than-give-away-its-covid-vaccine-oxford-makes-a-deal-with-drugmaker/.

24. David Meyer, "AstraZeneca Is Finding It Hard to Keep Its COVID-19 Vaccine Promises," *Fortune*, January 25, 2021.

25. Simon Allison, "Bill Gates, Big Pharma and Entrenching the Vaccine Apartheid," *Mail and Guardian*, January 30, 2021, https://mg.co.za/health/2021-01-30-bill-gates-big-pharma-and-entrenching-the-vaccine-apartheid/.

26. "Serum Institute of India Obtains Emergency Use Authorisation in India

for AstraZeneca's COVID-19 Vaccine," AstraZeneca, January 6, 2021, https://www.astrazeneca.com/media-centre/press-releases/2021/serum-institute-of-india-obtains-emergency-use-authorisation-in-india-for-astrazenecas-covid-19-vaccine.html#!.

27. Lauren Frayer, "The World's Largest Vaccine Maker Took a Multimillion Dollar Pandemic Gamble," *Goats and Soda* (blog), NPR, March 18, 2021, https://www.npr.org/sections/goatsandsoda/2021/03/18/978065736/indias-role-in-covid-19-vaccine-production-is-getting-even-bigger.

28. Gavi Alliance Staff, "New Collaboration Makes Further 100 Million Doses of COVID-19 Vaccine Available to Low- and Middle-Income Countries," Gavi.org, September 29, 2020, https://www.gavi.org/news/media-room/new-collaboration-makes-further-100-million-doses-covid-19-vaccine-available-low.

29. Tulip Mazumdar, "India's COVID Crisis Hits Covax Vaccine-Sharing Scheme," *BBC News*, May 17, 2021.

30. Sara Jerving, "Serum Institute of India Resumes COVAX Shipments of COVID-19 Vaccines," *Devex*, November 26, 2021, https://www.devex.com/news/serum-institute-of-india-resumes-covax-shipments-of-covid-19-vaccines-102199.

31. Angus Liu, "With 200M Unused Doses, AstraZeneca's COVID Vaccine Partner Serum Institute Halts Production," *Fierce Pharma*, April 22, 2022, https://www.fiercepharma.com/pharma/200m-unused-doses-astrazenecas-covid-vaccine-partner-serum-institute-halts-production.

32. Bryce Baschuk, "WTO Approves Vaccine-Patent Waiver to Combat COVID Pandemic," *Bloomberg*, June 16, 2022, https://www.bloomberg.com/news/articles/2022-06-17/wto-approves-vaccine-patent-waiver-to-help-combat-covid-pandemic.

33. Kevin Dunleavy, "PhRMA Says COVID-19 Vaccine Patent Waiver Is a 'Political Stunt,' While Advocate Argues It Doesn't Go Far Enough," *Fierce Pharma*, June 21, 2022, https://www.fiercepharma.com/pharma/wtos-patent-waiver-covid-vaccines-political-stunt-phrma-says.

34. Svet Lustig Vijay and Elaine Ruth Fletcher, "Gates Foundation: Technology Transfer, Not Patents Is Main Roadblock to Expanding Vaccine Production," *Health Policy Watch*, April 14, 2021, https://healthpolicy-watch.news/patents-are-not-main-roadblock-to-expanding-vaccine-production-says-top-gates-foundation-official/.

35. Agence France-Press, "South Africa Hails COVID-19 Vaccine Patent Waiver," *VOA News*, June 18, 2022, https://www.voanews.com/a/south-africa-hails-covid-19-vaccine-patent-waiver/6623167.html.

36. Alexander Winning and Joe Bavier, "African Union Says Russia Offers 300 Million Doses of Sputnik V Vaccine," *Reuters*, February 19, 2021.

37. Michelle Fleury, "Bill and Melinda Gates Divorce after 27 Years of Marriage," *BBC*, May 4, 2021.

38. Halleluya Hadero and Associated Press, "Gates Foundation Brings On New Four Board Members following Divorce," *Fortune*, January 26, 2022, https://fortune.com/2022/01/26/gates-foundation-board-expansion-divorce/; Emily Glazer, "Gates Foundation Adds Board Members after Billionaires' Divorce," *Wall Street Journal*, January 26, 2022, https://www.wsj.com/articles/gates-foundation-adds-board-members-after-billionaires-divorce-11643205603?mod=article_inline.

39. Emily Glazer, "Melinda French Gates No Longer Pledges Bulk of Her Wealth to Gates Foundation," *Wall Street Journal*, February 2, 2022, https://www.wsj.com/articles/melinda-french-gates-no-longer-pledges-bulk-of-her-wealth-to-gates-foundation-11643808602.

40. Nicholas Kulish, "What the Gates Divorce Means for the Bill and Melinda Gates Foundation," *New York Times*, May 4, 2021, https://www.nytimes.com/2021/05/04/business/bill-melinda-gates-divorce-foundation.html.

41. Bill Gates, *How to Prevent the Next Pandemic* (New York: Alfred A. Knopf, 2022).

42. Bill Gates, "3 Things We Can Do Right Now," *GatesNotes* (blog), April 30, 2022, https://www.gatesnotes.com/Health/3-things-we-can-do-right-now-to-prevent-the-next-pandemic.

43. Gates, *How to Prevent the Next Pandemic*, 17.

44. Gates, 22.

Index

2008 economic crisis, 3–4

Africa, 4–5, 35–36, 45–46, 76–81, 86, 103, 116, 124, 140, 148, 152–53, 156–57, 161, 180n13
Africa Centres for Disease Control and Prevention, 156
African Union, 157
aid agencies, 1–2, 8
Alliance for Progress, 30
Alma-Ata Conference (1978), 7–8, 70–71, 74–75, 84–85, 107, 125, 130, 134, 148–49
Amin, Samir, 35, 103
apartheid, 11, 36–37, 77, 86, 112, 114–17, 120–21, 124–29, 131, 142, 145, 197n62
Asia, 4, 17, 36, 76, 78, 81, 95, 103, 106, 140–41, 180n13
AstraZeneca, 155–56
Atoms for Peace Conference (1955), 60
austerity, 6, 123, 151

Bandung Conference, 36, 81, 180n13
Bangkok, 140–41
Belgrade Conference, 36, 180n13
Biden, Joseph R., 153, 156
bioethics, 7
Blair pump, 114, 117
Blair toilet, 114, 117
Brand, Stewart, 9, 44, 48–49, 51, 91
Brazil, 63, 91, 106
Bretton Woods, 18
Brexit, 3–4
Buffett, Warren, 139, 157
Bunn, Tony, 144–46, 162
Bureau of Science and Technology (USAID), 109
Burma, 10, 13–17, 19–22, 25, 30, 34, 36–37, 43, 46, 49–50, 68, 142

California, 48, 51, 106, 152
CalTech, 61–62
cancer, 37, 120, 129, 146
capitalism, 6, 16, 20, 22–23, 26, 33, 56, 86, 98, 150–51, 166n19
Carter, Jimmy, 9, 34, 57, 59, 62, 71, 87–88, 94, 107
Centre of Science for Villages in Wardha, India, 103
Centro Intercultural de Documentación (CIDOC), 30, 32
Césaire, Aimé, 5, 79
Christian Science Monitor, 104–5
civil rights, 4, 52, 59
civil rights movement, 4, 59, 180n10
Cluett, Suzanne, 136
Cold War, 3, 6–8, 14, 17, 24–25, 28, 30, 36–37, 49, 81, 157, 172n3, 181n16
communism, 16, 119
community health, 8, 37, 64, 84, 124, 129–32, 144
Conference of Ministers of African States Responsible for the Application of Science and Technology to Development (CASTAFRICA), 74, 77–81, 90, 116, 121
COVAX, 154–56
COVID-19 pandemic, 3, 11, 143, 150–58

Cuba, 36–37, 132, 146
cultural hegemony, 4
cybernetics, 49, 58, 177n49

dams, 2, 13, 58
decolonization, 4, 78
Democratic Republic of the Congo, 1, 3, 152, 200n32
developing countries, 8, 24, 27, 35, 41, 44, 61, 63, 65–69, 73–74, 80–81, 83–87, 89, 95, 97–98, 100–1, 114, 122, 182n32, 184n68, 195n25
development: aid, 92, 108, 122; alternatives to, 31; capitalist, 21, 68, 89; community, 22, 25–26, 31, 33, 39, 56, 75, 115; economic, 5, 13, 16, 20–21, 24–26, 28, 30, 33, 39, 57, 123; economists, 36, 61, 77, 122; experts, 31, 36, 64; incremental, 15; industrial, 21, 46; and modernization, 5, 8, 111; needs, 43; networks, 41; and NGOs, 101; and PATH, 99–100, 106, 110; pharmaceutical, 70–71, 84, 88, 95–96, 111, 136, 154; planners, 2, 14, 40; policy, 7, 12, 14, 22, 83, 90, 122–26, 147; practitioners, 40, 45, 142; projects, 10, 22, 88, 108–10, 115–16, 121–22, 145–47; rural, 25, 121; small-scale, 34; solution, 76, 116; and the Soviet Union, 14, 21; sustainable, 3, 143, 165n2; technicians, 10, 39; and technology, 2, 11, 17, 22, 27, 41, 58–60, 62–63, 68, 81, 88, 93–95, 106–7, 116, 121, 133, 135, 139, 143, 162; and the United States, 3, 14, 21, 25, 45, 59, 73, 83, 88–89, 97; WASH, 93; Western, 14, 17, 126; and women, 102, 113–15, 121. *See also* international development
Discovery Center, 134–35, 140
disease, 64, 69–70, 108, 110–11, 120, 125, 130, 135, 143, 146, 148, 152–53, 156, 175n9, 195n25; campaigns, 72–73, 82, 109; control, 71, 84, 108; iatrogenic, 31–32; patterns, 72–73; vertical, 69
drugs, 1–2, 32, 94, 111, 152, 157, 175n9, 189n68, 200n20; accessibility of, 69, 73, 142, 147; essential, 8, 59, 70–72, 74, 84–85, 107, 120, 167n28, 183n59, 187n44, 193n66; HIV/AIDS, 138, 148; manufacturing of, 139
Duncan, Gordon, 91, 95–96, 98

Eastern Cape Appropriate Technology Unit (ECATU), 129, 143
Ebola, 3, 143, 200n32
Economic and Social Board, 12–14
economics, 4, 10, 12, 16–23, 28–29, 33, 35, 56, 75–76, 114, 134; Buddhist, 15–17, 26, 34, 49; Gandhian, 23
Egypt, 5, 36, 63, 69, 180n13, 193n66
Eisenhower, Dwight D., 59
Elias, Chris, 140–42, 156, 200n31
environment, 2, 21, 50, 56, 102
environmentalism, 10, 31, 51, 56, 92
European Union, 3–4, 29

Fanon, Frantz, 5, 86
feminism, 10, 40, 52–56, 118, 121, 178n69, 179nn74–77; second-wave, 4
Ford Foundation, 9, 25, 47, 81, 91, 95, 98–99, 101, 106–7, 122, 148, 191n26
foreign aid: and Africa, 35; benefits of, 17; bilateral, 66, 68, 115; budgets, 43, 57, 64, 90; dependency on, 53; investments, 24; and manufacturing, 5; and modernization theory, 174n47; organizations, 40, 115, 121; and Point Four program, 8; and the private sector, 2, 66, 89, 92; programs, 11, 93, 96, 100, 111–12, 142, 176n22; and public support, 5, 58–59, 66, 70, 72; and technology, 7, 58, 147; and VITA, 42
Foreign Assistance Act (1961), 59
Fritz, Dale, 39, 41–42, 44

Gandhi, Mohandas K., 10, 16, 19, 22–26, 32, 39, 94, 103, 113, 171nn60–61
Gates, Bill, 9, 91, 95, 134–39, 146, 153–55, 157–58, 199n8
Gates, Melinda, 134, 137, 157
Gates Foundation, 9, 135–41, 143–45, 153–57, 200n22
Geneva, 61, 75–76, 84, 114, 148
Ghana, 5, 36, 46, 81, 86, 144, 176n32
Giridharadas, Anand, 137
global capitalism, 56, 86, 150

global health, 2–3, 7–10, 92, 111, 133, 137, 139–41, 143, 145–48, 150–51, 153, 156–58, 160, 162–63, 167n19, 183n58, 189n68, 189n9, 193n9; programs, 9, 11, 94, 100
Global Health Innovation Accelerator (GHIA), 133, 143–45, 153
globalization, 4, 21, 105
Global North, 3, 31, 75, 93, 111, 133
Global South, 7, 9, 31, 35, 75–76, 92–94, 123, 133, 135, 145, 149, 155, 167n22, 189n68, 195n25, 201n50
global subsistence rights, 83
Global Vaccine Alliance (Gavi), 137–40, 154–55
GOBI (growth monitoring, oral rehydration, breastfeeding, immunization), 8, 74, 107, 109
Goma, 1
Gordon, Richard, 145, 162
green revolution, 114
Grimstad, Kirsten, 52–53, 55–56
Guardian, 4

Harbison, Frederick, 61
Harley, Richard, 105
Harvard, 48, 123, 126, 138
health care, 37–38, 94, 112, 123, 133, 143, 148, 153–54, 162, 179n75; primary, 7–8, 11, 59, 64, 69–75, 82, 84–85, 107–11, 115–16, 119, 124–25, 129–32, 134, 149, 151, 183n61, 185n71; sector, 72, 83, 103
health education, 84
Hesburgh, Theodore, 58–60, 62–63, 87
highways, 2, 13–14, 58
Hinduism, 10, 22–23, 27
HIV/AIDS, 111, 131–32, 138, 146, 148, 153
humanitarianism, 1, 8–9, 103, 167n28
human needs, 22, 41, 59, 64–67, 70–71, 74–75, 82–83
hunger, 51, 64

immunization, 8, 74, 99, 109
India, 2, 5, 10, 22–25, 34, 36, 46, 49, 68, 98, 100, 103, 113, 121, 142, 155–56, 180n13, 197n54
indigenous industry, 5, 22, 31, 76, 122, 129, 133, 153, 156
Indonesia, 5, 36, 109, 180n13
inequality, 51
infrastructure, 3, 8, 14, 39, 43, 63, 73, 92, 97, 111–12, 135, 148, 152; projects, 2, 9, 13
Institute for Scientific and Technical Cooperation (ISTC), 63, 71, 73, 88–89
Institute of Medicine conference, 71–72, 88
Intermediate Technology Development Group (ITDG), 29, 32, 40, 43–49, 56, 85, 103–5, 115, 122, 142, 147–48, 161, 176n26
International Conference on Primary Health Care (WHO). *See* Alma-Ata Conference
international currency, 18
international development, 1–2, 8, 10, 26, 33–34, 39, 41, 47, 49, 56, 57, 60, 72, 82–83, 102, 129, 171n68
International Development Cooperation Agency (IDCA), 88–89
international health, 2–3, 7–8, 92; campaigns, 10, 56, 74, 82; and NGOs, 90, 94, 108, 136, 148; policy, 71–72, 82, 84, 143, 189n69; and technology, 75, 114, 116; and the United States, 165n3
International Institute for Rural Reconstruction, 44
International Science and Technology Institute, 67
international volunteerism, 41, 175n2
interventionism, 7, 70

Johnson, Lyndon B., 42, 53, 82

Kennedy, John F., 5, 59, 69–71
Keynes, John Maynard, 18, 32–33, 82, 170n39, 173n32
khaddar, 16
Kivu, Lake, 1
Kohr, Leopold, 28–31, 34, 37–38, 172n3
Kumasi, 46–47, 176n35
Kumasi Technology Group, 46

latrines, 8, 25, 39, 41, 117
libertarianism, 33, 51, 178n64

Mahler, Halfdan, 75, 82, 84, 107, 148–49
Mahoney, Richard, 91, 95, 97–101, 136–37
Manhattan Project, 61
manufacturing, 2, 5, 14, 58, 63, 69, 83, 96–98, 111, 121, 138, 150, 153–56
Maputo Conference on Health in South Africa (1990), 131
Marxism, 18, 20, 28, 32, 35, 76, 103
Massachusetts Institute Technology (MIT), 48, 61–62, 181n17
materialism, 14–16, 20, 58
McDermott, Walsh, 60–61, 70, 184n62
McNamara, Robert, 64, 82–83
medicine, 8, 20, 31–32, 38, 54, 57, 70–72, 97, 109, 117–20, 125, 146, 173n17, 195n25, 202n7
Memmi, Albert, 5
Microsoft, 95, 111, 135, 137–38
Millikan, Max, 61, 181n17
mining, 21, 67, 119
model village, 24–25, 29, 43, 68, 96, 103, 113, 116–17, 121, 123, 127, 183n56
modernity, 58, 116, 133, 195n25
modernization, 5, 15, 22, 27, 57–58, 64, 72–73; projects, 2, 8, 43, 59, 82; scientific, 24
modernization theory, 2, 13, 58, 64, 111, 169n11, 174n47, 181n17
Morgan, Peter, 117–18
Mugabe, Robert, 9, 113, 116–17, 120–21, 123, 129, 194n3, 195n23
Mujuru, Joice, 11, 113–14, 121, 133, 194n3

Nairobi, 93–94, 122, 141, 147
Nasser, Gamal Abdel, 5, 36
National Academy of Medicine, 69
National Academy of Sciences, 60–61
nationalism, 5
National Science Board, 59
National Science Foundation, 62, 104
National Training Centre for Rural Women, 113–14
Nehru, Jawaharlal, 5, 22, 24, 26, 36, 113, 171n60
neocolonialism, 7, 86, 180n13
neoliberalism, 6–7, 56, 64, 75, 92, 95, 108, 122, 147, 151, 166n19, 199n19, 201n48
New Directions, 59, 66, 71
New International Economic Order (NIEO), 5, 59, 62, 65, 69, 74–77, 81–82, 85, 88–89, 115, 147
New York, 17, 24, 37, 39, 42, 55, 91, 101–2, 151, 153
NGO Forum on Science and Technology for Development, 102, 105–6
Nixon, Richard, 59
Nkrumah, Kwame, 5, 36, 47, 81, 86
Non-Aligned Movement, 5, 36–37, 81, 180n13
nongovernmental organizations (NGO), 8, 10, 90, 92, 94, 101–2, 105–6, 111–12, 116, 142, 148, 180n9, 183n48, 190n20
nonviolence, 20–22, 37
Nu, U, 12–15

Office of Science and Technology Policy, 62, 71, 89
oil shocks (1973), 5, 21, 34, 57–58, 64

Pax Americana, 3
Peace Corps, 30, 41–42, 44, 59, 136, 175n2, 175n11
Perkin, Gordon, 91, 95, 98–101, 136–37, 139–42, 162
pharmaceuticals, 14, 63, 65–66, 69–73, 84, 95, 98, 108–11, 116, 138, 155–56, 182n32, 184n68, 193n66. *See also* drugs
Planned Parenthood, 136
Point Four Program, 8, 24–25
Poonawalla, Adar, 156
Popular Science, 99
Population Council, 101, 140
poverty, 14, 25, 36, 51, 64, 71, 82, 135, 147; Black, 53; War on Poverty, 53; world, 56
power plants, 2, 13–14
Press, Frank, 61, 89
privatization, 6, 76, 92, 95, 107, 151
Program for Appropriate Technology in Health (PATH), 9–10, 92, 94–101, 106–8, 110–11, 133, 135–37, 139–46, 148, 155
Program for the Introduction and Adaptation of Contraceptive Technology (PIACT), 91–92, 95–96, 136, 142

public-private partnerships, 8, 94, 98, 100, 139, 153–55

Ranis, Gustav, 64–65, 71
Reagan, Ronald, 6, 56, 81, 107, 110, 178n69, 190n20
Reaganism, 6–7
Rennie, Susan, 52–53, 55–56
reverse innovation, 3, 144
Rockefeller Foundation, 59, 61, 70, 74, 121
Rostow, Walt W., 2, 13, 181n17
Rottenburg, Richard, 127
rural clinics, 8, 72
Russia, 112, 157; and technical aid, 15
Russian Tea Room, 55

Saké, 1
Schumacher, Ernst Friedrich, 4, 10, 12–23, 26–40, 44–46, 49, 51, 56, 58, 64, 83, 85, 93, 103–5, 115, 126, 142, 147, 161
science: access to, 5, 34; and Africa, 78–79; and the Cold War, 181n16; and colonial rule, 5; and development, 58, 60, 62, 71, 102, 111, 116, 139; and economics, 15, 20; history of, 160; modern, 24; and natural resources, 21; and NGOs, 103; policy, 60, 89, 180n12; and religion, 19; and the Soviet Union, 61; and technical knowledge, 46; and technology studies, 7, 118, 175n, 176n20; and the Third World, 60; transfer, 37, 74; and the United States, 61, 63, 81, 87, 89–90; and vaccines, 142; and Whole Earth, 50; and women, 94; and Zimbabwe, 120
Science Advisory Committee, 60
Seattle, 92, 95, 101, 111, 134, 136, 140, 143, 153
Silicon Valley, 41, 49, 51, 91–92, 95, 143, 147, 150, 192n60
Small is Beautiful: Economics as If People Mattered (1973), 4, 17, 32–34, 47–49, 56–57, 101, 171n64, 174n44
socialism, 5, 18, 23, 26, 33–34
social medicine, 8, 125, 130
solar panels, 1–2, 103
South Africa, 2, 9, 36–37, 77, 86, 112, 114–16, 119–21, 124, 126–33, 142–46, 152–53, 155–56; National Progressive Primary Health Care Network, 11
South African Medical Research Council (SAMRC), 143–46, 153, 162
Strategic Health Innovation Programme (SHIP), 145
Subcommittee on Domestic and International Scientific Planning, Analysis, and Coordination of the Committee on Science and Technology, 64
Sukarno (Koesno Sosrodihardjo), 5, 36
swadeshi, 16, 22, 197n54
Syria, 3

technical assistance, 6, 8, 13, 39, 88, 122
Technologies for Primary Health Care (PRITECH), 107–10
technology: access to, 152; appropriate, 2–12, 27, 31, 34–35, 38, 40–41, 44, 46–47, 49, 54, 56–59, 62–64, 66–69, 71, 73–76, 80–81–85, 89–90, 92–94, 96–97, 99–104, 106, 110–13, 115–18, 120–21, 123–27, 129–31, 133, 135, 138–44, 147–48, 151, 165n2, 168n32, 171n61, 185n1, 196n34; Bantu, 127; capital-intensive, 27, 35, 68, 123; and capitalism, 21; and the Cold War, 24; and colonial rule, 5, 53; contraceptive, 101; control over, 54–55; and development, 14, 17, 21–22, 41, 58, 60, 62, 83, 95, 103, 107, 111; and disaster, 28; disruptive, 3, 7; dissemination of, 5; and the environment, 28, 38, 49; and globalization, 105; health, 8, 84, 100, 110, 133, 139, 145–47, 149, 152; high, 35, 60, 89, 118; history of, 6–7, 176n20; and imperialism, 35; and India, 197n54; and individuation, 3; industrial, 66; inhuman, 34; innovation, 147; intermediate, 2, 26–27, 33, 38, 45–46, 83, 173n38; justice, 147; labor-saving, 21, 114; and limits, 50; and Luddism, 40; and materiality, 7; mobile, 135; modern, 15, 17, 24, 27; and Mumford, Lewis, 24; and natural resources, 21, 63; point of use, 2, 11; and Schumacher, Ernst Friedrich, 31, 34–35, 56, 142; selection, 32, 58;

technology (*continued*)
simple, 2; small-scale, 10, 26, 47; soft, 49, 140; and the Soviet Union, 61; space, 61; studies, 7, 175n8, 176n20; systems of, 7; traditional, 27; transfer, 5, 35, 37, 57–58, 65–66, 68, 72–75, 78, 81, 86–88, 98, 103, 108, 110, 155–56; underdeveloped, 35; and USAID, 59, 109; vernacular, 31; Western, 27; and women, 10; and work, 25, 70
Technology Consultancy Centre, 47
Tet Offensive, 5, 43, 58
Thailand, 44
Thant, U, 12–13, 36–37, 60, 174n55
Thatcher, Margaret, 6
Third World, 53, 60, 67, 71, 82, 90, 102, 104, 121, 123, 169n20
Tito, Josip Broz, 5, 36
Truman, Harry, 8
Trump, Donald J., 153–54, 156, 200n22

Ukraine, 3
UN Conference on the Application of Science and Technology for the Benefit of the Less Developed Countries (UNCAST), 60–62, 70, 89–90, 180n13
UN Conference on Human Settlements, 104
UN Conference of Science and Technology for Development (UNCSTD), 58–60, 62–68, 72–73, 86, 88, 101–2, 105–6, 138, 182n32
UN Conference on Trade and Development (UNCTAD), 5, 81, 83, 87
UN Educational, Scientific, and Cultural Organization (UNESCO), 77–78, 81–82, 104–5, 121, 156
United Kingdom, 3, 10, 18, 27, 32, 56, 82, 118–19, 142, 146, 176n26, 196n34
United Nations, 9, 12–13, 30, 37, 59, 62, 68, 75, 77, 86, 87, 93, 105, 114, 119, 147, 182n32; organizations, 57, 78, 81, 104
United States Agency for International Development (USAID), 1; and appropriate technology, 57, 59, 67, 75; assistance program, 68–69; budget of, 5, 7, 43, 58, 88, 92; and donors, 131; and the Ford Foundation, 148; and foreign aid, 89; funding, 41–42, 63, 88–89; and Gates Foundation, 9; and health care, 71–73, 107, 109; and Levin, Sander, 66; and Mauritania, 67; and modernization, 59, 64; and Nepal, 136; and PATH, 100, 107, 140–41; and pharmaceuticals, 72; programming, 10, 34, 39, 59, 89–90, 109–10
University of California, Berkeley, 48
University of Science and Technology, 46
UN Technical Assistance Board, 13
UN Women's Conference (1975), 93–94, 114
US Foreign Service, 62
uterine balloon tamponade (UBT), 144

vaccines, 108, 150, 153–55; affordability, 139; coolers, 134; COVID-19, 154, 156–58; development, 88, 136; famine, 156; Global Vaccine Alliance, 137, 154; hepatitis B, 111; HIV, 153; labels, 97; meningitis A, 142; producers, 138, 155; programs, 138; vial monitor (VVM), 99, 135, 140, 142
Vienna Conference (1979), 58–59, 62, 67, 69, 72–74, 85–90, 147
Vietnam War, 5, 32, 37, 43, 58
Village Technology Center, 43–44
Village Technology Center Catalog, 44
Village Technology Handbook, 10, 39, 41, 43–45, 48, 53
violence, 1, 20, 93; domestic, 122; sexual, 129
Volunteers in Technical Assistance (VITA), 39–49, 53, 55–56, 61, 64, 87, 104–5, 121, 142, 148, 175n2, 175n10

Westernization, 14
Whole Earth Catalog, 10, 32, 40, 44, 48–49, 51–53, 55, 92, 104, 178n64
Whole Earth Epilog, 49
women: Black, 53, 114, 129, 190n25; of color, 4, 53–54, 93, 96, 178n69; and *Community Radio*, 132; and contraception, 96; and development, 102, 115–16, 121–22; empowerment, 54, 93; and feminism, 4; and India, 23; and international health campaigns, 10–11;

in Kenya, 147; and labor, 93–94; low-income, 123, 178n69; queer, 4, 53–54, 178n69, 179n74; and reproductive capacity, 52; rural, 113–14, 121; self-health, 54, 108; and sexism, 53; and technology, 10, 55, 93–94, 125–27, 130, 133, 162; and the Third World, 53; and the UN Decade for Women, 93, 114; and the United States, 52, 62, 96, 157; violence against, 93; white, 53; and *Whole Earth*, 50, 52–53
World Bank, 7, 9, 59, 64, 68, 72, 74, 82–84, 89, 111, 126, 163, 168n29
World Health Organization (WHO), 7, 9, 71, 74–76, 82, 84, 87, 89, 99–101, 107–11, 117, 120, 125, 140, 147–48, 154, 156, 162–63, 184n68
World Trade Organization (WTO), 75, 138–39, 156
World War II, 3–4, 29, 39, 55, 77, 79, 82, 174n47

Yemen, 3
Yugoslavia, 5, 36–37, 63

Zambia, 62, 119, 125
Zimbabwe, 7, 9, 11, 112–16, 119–23, 127, 133, 142, 147, 194n4
Zimbabwe Bush Pump Type B, 117–18, 168n32